The Business of Additive Manufacturing

Although additive manufacturing (AM), also known as 3D printing, has been around for almost 40 years, few people know how it actually works and the huge impact and benefits it offers. This book explains what AM is, using business theories to explain and illustrate why AM is increasingly being used across industries.

The book translates complex engineering technology into relevant managerial terminology, using real-world examples from industries such as apparel, construction and transportation. It provides an introduction into the technical background of AM before expanding on the applications, opportunities and challenges to business models.

Offering a unique managerial perspective, this book is aimed primarily at a scholarly audience and those researching across business disciplines, including technology management, manufacturing, production and operations management. It can also be used in emerging business courses on AM.

Harm-Jan Steenhuis is Professor of Management, International Business in the College of Business at Hawai'i Pacific University. He previously worked at Eastern Washington University, North Carolina State University, and the University of Twente in Enschede, The Netherlands. He received his MSc in Industrial Engineering and Management and his PhD in International Technology Transfer from the University of Twente.

His main research areas connect with technological development and economic development-related issues such as clusters, technology transfer and diffusion, and innovative technologies. In recent years, he has focused on additive manufacturing (AM) such as the diffusion of industrial AM, adoption of desktop 3D printers, and AM in different industries.

RIOT!

Routledge Studies in Innovation, Organizations and Technology

Advancing Big Data Analytics for Healthcare Service Delivery
Tiko Iyamu

Open Labs and Innovation Management
The Dynamics of Communities and Ecosystems
Edited by Valérie Mérindol and David W. Versailles

Technology Brands in the Digital Economy
Edited by Wioleta Kucharska and Ewa Lechman

Business Models for Industry 4.0
Concepts and Challenges in SME Organizations
Sandra Grabowska and Sebastian Saniuk

Innovation in the Digital Economy
New Approaches to Management for Industry 5.0
Edited by Agnieszka Rzepka

Information Technology in Contemporary Organizations
Redefining IT Management for Organizational Reliability
Katarzyna Tworek

Impact of Artificial Intelligence in Business and Society
Opportunities and Challenges
Davide La Torre, Francesco Paolo Appio, Hatem Masri, Francesca Lazzeri and Francesco Schiavone

Management System for Strategic Innovation
Building Dynamic Capabilities View of the Firm
Mitsuru Kodama

For more information about this series, please visit: www.routledge.com/Routledge-Studies-in-Innovation-Organizations-and-Technology/book-series/RIOT

The Business of Additive Manufacturing

3D Printing and the 4th Industrial Revolution

Harm-Jan Steenhuis

Routledge
Taylor & Francis Group
LONDON AND NEW YORK

First published 2024
by Routledge
4 Park Square, Milton Park, Abingdon, Oxon OX14 4RN, UK

and by Routledge
605 Third Avenue, New York, NY 10158, USA

Routledge is an imprint of the Taylor & Francis Group, an informa business.

British Library Cataloguing-in-Publication Data
A catalogue record for this book is available from the British Library.

ISBN: 978-1-032-50572-5 (hbk)
ISBN: 978-1-032-50573-2 (pbk)
ISBN: 978-1-003-39910-0 (ebk)

DOI: 10.4324/9781003399100

Typeset in Sabon
by SPi Technologies India Pvt Ltd (Straive)

Contents

1 Manufacturing in the 21st century

As the 20th century came to an end and the 21st century started, the general notion has been that manufacturing is considered less and less important, especially for developed countries. Questions were raised about the future of manufacturing, and the use of tangible resources to make physical products in factories was no longer considered a viable way to characterize manufacturing (Fleury, Gregory and Bennett, 2007). Two main reasons can be identified for this perception of a manufacturing decline: services and developing nations.

Shift to services

Advanced industrial nations started to shift more and more towards economies that relied primarily on services. Figure 1.1 illustrates how services have become more important compared with manufacturing for France.

Worldwide, the manufacturing value added as a percentage of gross domestic product (GDP) has also declined slightly over the last 20 years and is around 16% (see Figure 1.2).

The US and the UK are often used to illustrate the decline of manufacturing in industrialized countries. These countries, as well as several other industrialized nations, have indeed experienced manufacturing declines in terms of importance to their overall GDP, as can be seen in Figure 1.3.

What is underestimated is that services have tangible elements (Sampson, 2001) that require manufacturing: hotels need beds, restaurants need utensils, and airlines need airplanes. Thus, an expansion of services is likely to expand manufacturing as well. Also, in some countries, manufacturing still represents roughly 20% or more of the GDP. This includes industrialized nations such as South Korea, Germany and Japan (see Figure 1.4). This includes only countries that have at least roughly $50 billion value added in manufacturing, so small countries that may have a disproportionate share in manufacturing are excluded.

DOI: 10.4324/9781003399100-1

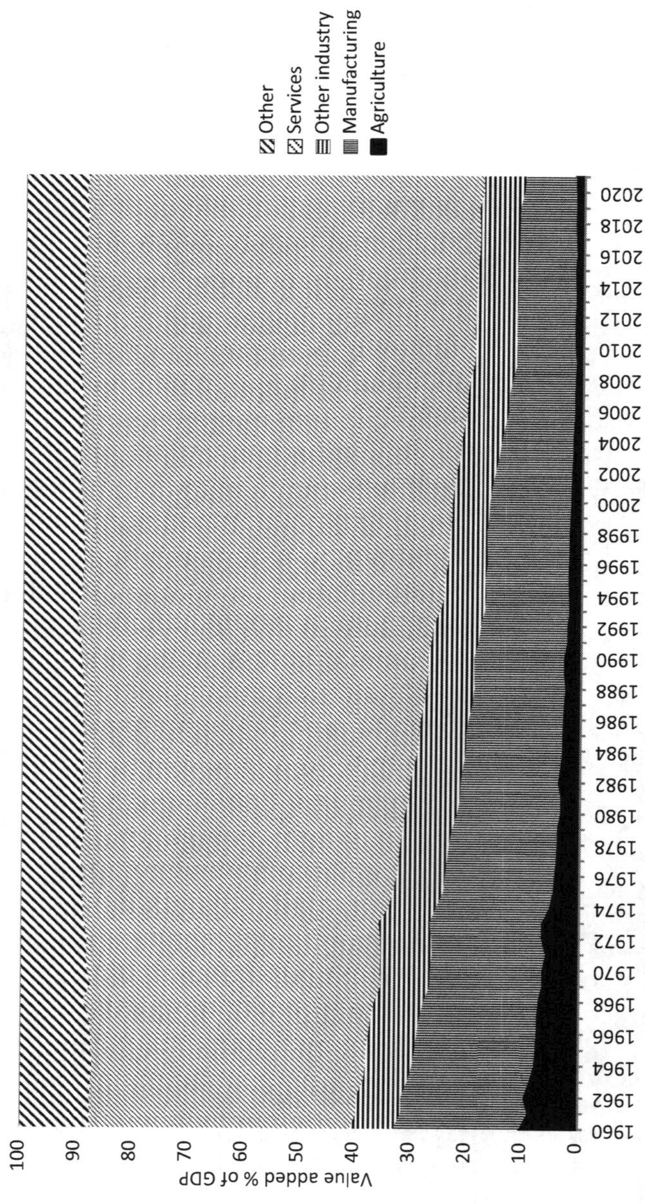

Figure 1.1 France's shift towards more service.

Source: adapted from the World Bank (www.data.worldbank.org)

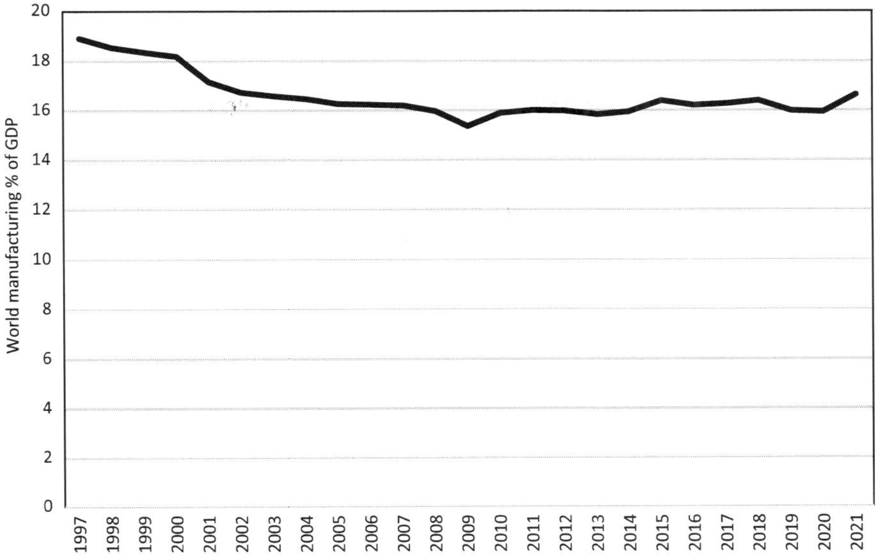

Figure 1.2 Worldwide manufacturing value added (percentage of GDP).
Source: adapted from the World Bank (www.data.worldbank.org)

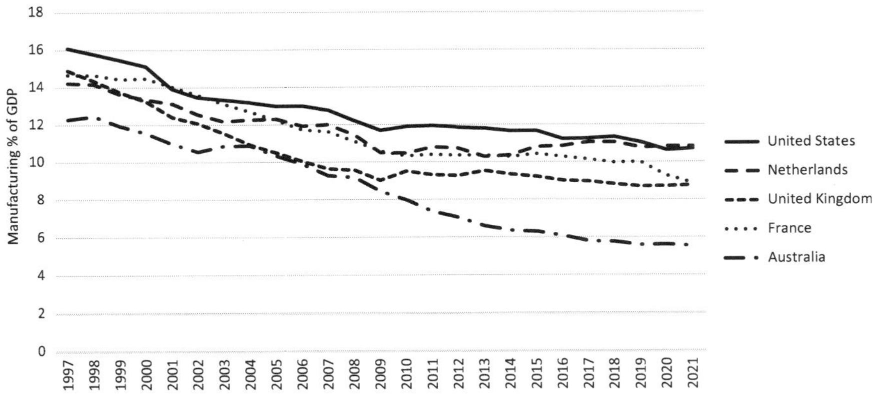

Figure 1.3 Industrialized nations with declining importance of manufacturing.
Source: adapted from the World Bank (www.data.worldbank.org)

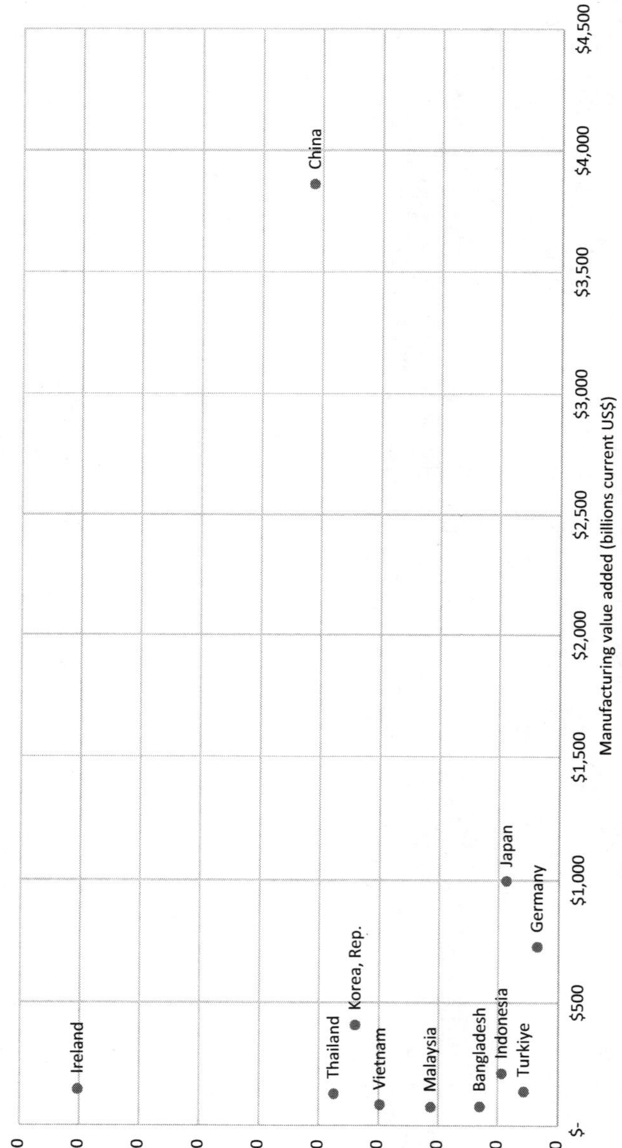

Figure 1.4 Select countries' manufacturing value added in 2020.

Source: adapted from the World Bank (www.data.worldbank.org)

Rise of developing countries

A second reason for the perceived decline in manufacturing is growth of manufacturing in developing nations. Figure 1.4 illustrates how China has become a manufacturing powerhouse and this is often perceived as coming at the expense of developed nations. There is a long history of research on the (international) location of manufacturing which shows this isn't necessarily the case. Ferdows (1997) showed that companies moved manufacturing to foreign factories for three primary reasons: access to low-cost production, access to skills and knowledge, and proximity to market. While cost gets a lot of (media) attention, the proximity to the market is the most frequently cited and most stable reason to have a plant in a foreign location (Vereecke and Van Dierdonck, 2002). Cost is an unstable factor that may be related to reshoring. Reshoring is moving manufacturing back to the home (often a developed) country. It occurs due to unexpected, rising, or hidden and greater-than-initially-expected cost (Benstead, Stevenson and Hendry, 2017). A study on outsourcing decisions found that companies generally do not comprehensively measure the costs of global sourcing and significantly underestimate the true costs incurred (Platts and Song, 2010). Exchange rates and rising labor and energy costs can also change the situation (Gray, Esenduran, Rungtusanatham and Skowronski, 2017).

Reality of manufacturing

Graphs such as Figure 1.3 are distorting because they show manufacturing as relative in terms of GDP while the GDP itself grows as well. Figure 1.5 shows the absolute numbers for the US over the last 25 years and this illustrates that manufacturing has grown substantially.

Figure 1.6 illustrates the same for several other countries, including those from Figure 1.3. While manufacturing may have become *relatively* less important in these economies, in *absolute* terms, it has been growing. Manufacturing is more important than often perceived and is likely to become even more important because of the latest industrial revolution.

Industry 4.0

The world has experienced several industrial revolutions which have led to great improvements in economic prosperity. Previous industrial revolutions can be thought of as simultaneous worldwide events but this is not correct. Not all countries entered previous industrial revolutions at the same time, and some countries may still be working through earlier industrial revolutions (Stearns, 2013). Parts of the world are currently experiencing the fourth industrial revolution, or Industry 4.0. This latest industrial revolution is fundamentally different from the previous three industrial revolutions. The first three dealt with improved technologies, but Industry 4.0 deals with a fusion

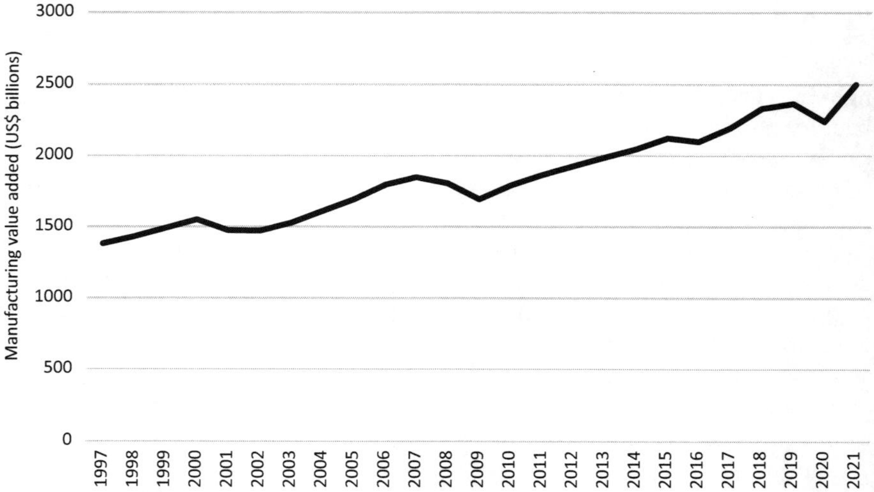

Figure 1.5 US manufacturing value added.

Source: adapted from the World Bank (www.data.worldbank.org)

of technologies and their interaction across the physical, digital and biological domains (Schwab, 2016) such as cyber-physical systems (CPSs) (Pereira and Romero, 2017). Fundamental concepts in CPSs associated with the virtual environment are Internet of Things, big data and cloud computing, whereas in the physical environment, it includes autonomous robots and additive manufacturing (AM) (Dilberoglu et al., 2017). Table 1.1 provides an overview of the four recognized industrial revolutions (Dosi and Galambos, 2013; von Tunzelman, 2003) with the related technologies.

Before the first industrial revolution, manufacturing was dispersed and took place on a small scale (e.g., a blacksmith). The first industrial revolution introduced new technology with economies of scale that caused the concentration and organization of manufacturing in factories. The second and third industrial revolutions introduced newer technologies with increasing productivity levels and levels of concentration which kept factories in place. The fourth industrial revolution is now (partially) reversing concentration of activities because of AM. While previous technologies required large scale, AM is a technology that is productive at low production volumes and is leading to increased dispersion of manufacturing.

Additive manufacturing

With AM, a product is made layer by layer. This will be explained in detail in Chapter 2. This process has several advantages, such as limited need for tooling, less waste, ability to manufacture complex and moving parts, and ability

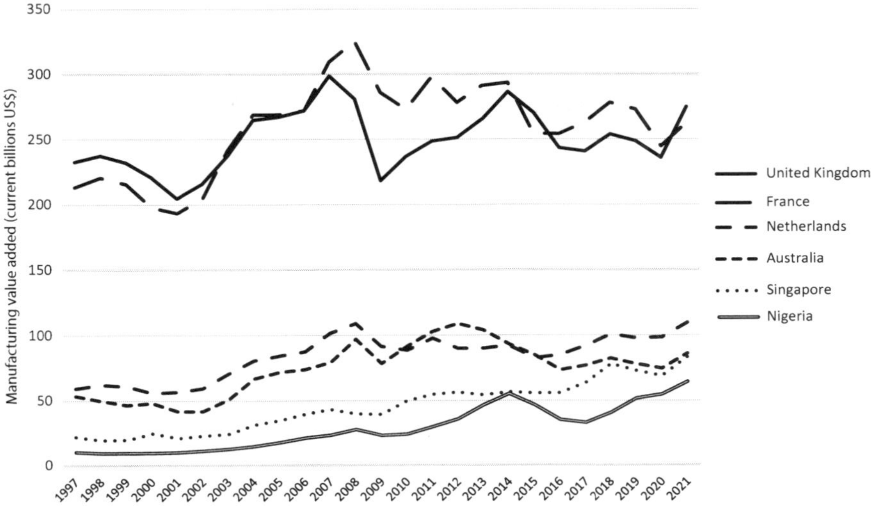

Figure 1.6 Manufacturing value added of a few selected countries.
Source: adapted from the World Bank (www.data.worldbank.org)

Table 1.1 Industrial revolutions

	First industrial revolution	*Second industrial revolution*	*Third industrial revolution*	*Fourth industrial revolution*
Approximate dates	1750–1815	1870–1914	1973–	2015–
Location	UK	USA, Germany	USA, East Asia	Many
Technologies	Steam	Electricity	Information and Communication Technology (ICT), biotechnology	Fusion of technologies across the physical, digital and biological domains

Source: adapted from Schwab, 2016 and von Tunzelman, 2003.

to manufacture closer to the point of consumption, thereby reducing logistics cost. The result of AM is that manufacturing in the 21st century has seen substantial changes affecting the cost of manufacturing and the profitability of manufacturing companies. It also affects in what type of locations manufacturing is feasible. This, therefore, relates to the ability of advanced industrial nations to remain competitive in manufacturing. AM might also serve as an enabler of reshoring of manufacturing (Fratocchi, 2018). The potential of

AM was recognized by the US when former US president Barack Obama mentioned in his 2013 State of the Union address the first manufacturing innovation institute in Youngstown, Ohio which is working on AM.[1]

AM is also known as 3D printing (3DP), and in this book both terms will be used interchangeably. Unfortunately, neither term is particularly effective in communicating the capabilities to a general audience. The more technical term, AM, relates to the types of additive processes that are followed which is not at all tailored towards general audiences to help them understand what AM actually is, whereas '3D printing' conjures up images of a printer/copier that somehow works in three dimensions (e.g., with three-dimensional layers of ink). At best, this can give people impressions of 3DP capabilities such as the ability to make plastic items that are a few inches high. The capabilities of AM are much more than what people typically envision. It has been used at nano-scale precision manufacturing but also for constructing buildings. AM has been used for cars, batteries, electronics, medical applications, and jewelry and has been applied in space. It has been used in the movie industry for props and backgrounds[2] and to help with animation[3] as well as for making furniture,[4] backpacks,[5] speed skating blades for Olympic gold medalists[6] and for recreating the art of Vincent van Gogh.[7] While it is not always easy to distinguish between the hype and reality of a new technology such as 3DP (Dedehayir and Steinert, 2016), it is clear that AM will have a huge impact on the world (Hornick, 2015).

Structure of the book

The purpose of this book is to educate people on this fascinating technology. This is done by providing a bridge of the technological (engineering) aspects and the (business) applications. AM and how it affects industries and our daily lives are explained. The structure of the book is depicted in Figure 1.7. The book consists of three parts and an introduction and conclusion. The first part of the book (Chapters 2–5) provides foundational information on AM. AM is described (*what*) and *how* it works is explained in Chapter 2. Initial business aspects are introduced in Chapter 3 to explain *why* AM is attractive. Chapter 4 examines the increasing use of AM by delving into common patterns of innovation diffusion. These types of patterns help to explain barriers to further growth. This leads to Chapter 5, which delves deeper into *where* it is advantageous to use AM.

The second part of the book (Chapters 6–10) delves into the actual use of AM. There are many broad examples of AM applications that speak to the imagination, so that choices had to be made on what to cover. Broadly speaking, Chapters 6 to 10 cover fashion, houses, food, healthcare, and transport. The discussion in each chapter is along the foundation chapters: What is 3D printed? (Practical examples are provided.) Why? Where? And further diffusion.

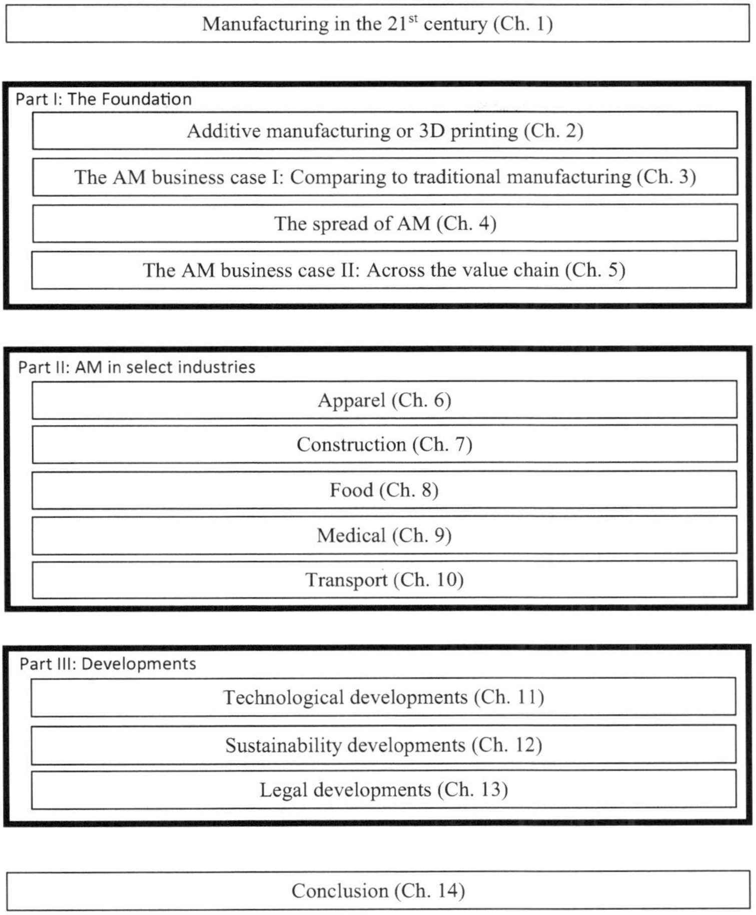

Figure 1.7 Structure of the book.

Source: author.

The third part (Chapters 11–13) takes a look at developments that are taking place and that can provide insight into what may be coming next. These developments are organized as technological (Chapter 11), sustainability (Chapter 12) and legal (Chapter 13).

Notes

1 https://obamawhitehouse.archives.gov/the-press-office/2013/02/12/remarks-president-state-union-address#annotations:15650729
2 https://www.voxeljet.com/case-studies/art-and-design/3d-printing-the-future-of-prop-manufacturing/

3 https://3dprintingindustry.com/news/lead-fabricator-reveals-how-3d-printing-helped-animate-netflixs-new-pinocchio-movie-219394/
4 https://www.fabbaloo.com/news/herman-miller-and-knoll-merger-creates-3d-printing-opportunities
5 https://www.3dnatives.com/en/3d-printing-aerorise-backpack-051220224/
6 https://3dprintingindustry.com/news/chinas-skate-team-takes-gold-at-winter-olympics-using-farsoon-3d-printing-technology-205155/
7 https://3dprintingindustry.com/news/ai-and-3d-printing-used-to-recreate-lost-van-gogh-artwork-in-ucl-initiative-215454/

Part I

The foundation

2 Additive manufacturing or 3D printing

Seven types

The purpose of this chapter is to explain how AM differs from the 'traditional manufacturing' (TM) methods. The seven different types of AM are distinguished and discussed.

Comparing 'traditional' and AM

Manufacturing is an industrial activity that changes the form of raw materials to create products (Ostwald and Muñoz, 1997: 1). AM is a process whereby a product is made layer by layer and is different from TM. There are several ways of categorizing TM processes, and a useful way is distinguishing forming, shaping, separating, bonding, and finishing. Forming is putting raw material in a shape (e.g., casting). Shaping is changing an earlier 'form' (e.g., by bending or forging). Separating involves, for instance, cutting and also drilling. Combining includes processes such as gluing, and finishing involves, for example, painting. The type of manufacturing process used is dependent upon the material characteristics. Thus, TM can be viewed as a combination of types of processes and materials as depicted in Table 2.1. In Table 2.1, the term 'polymers' is used. This is often used interchangeably with the term 'plastics'. Although not all polymers are plastics, all plastics are polymers. For the ease of the discussion in the remainder of this book, the term 'plastics' will generally be used.

AM is a different type of approach, and several of the traditional processes are sometimes combined into one AM process. For instance, if somebody wants to make a hollow metal cube, then with TM you start with two blocks of metal. For each block, metal would be cut away on a computer numerical control (CNC) machine so that it results in half of a hollow metal cube. These two halves would be combined by welding them together. With AM, a hollow object can be manufactured at once, and there is no need to have two parts.[1] In this instance, the AM approach would lead to less waste because there is no material that is getting cut away, and AM would also be a simpler process in the sense that it wouldn't require combining of two halves into one product. The more in-depth discussions in part I of the book will explain these advantages more.

DOI: 10.4324/9781003399100-3

Table 2.1 TM processes

		Manufacturing process				
		Forming	Shaping	Separating	Bonding	Finishing
	Metals					
Material	Polymers					
	Wood					
	...					

Source: author.

In terms of the history of AM, the technology has existed since 1980. In that year, a Japanese lawyer called Dr. Hideo Kodama, of the Nagoya Municipal Industrial Research Institute, was the first person to file a patent for Rapid Prototyping technology. However, the full application was not submitted on time, and he was not granted the patent.[2] There was also a group of three French inventors, but owing to a variety of reasons and a perceived lack of a potential market, they did not obtain a patent.[3] The initial patents ended up being granted by the US Patent and Trademark Office (USPTO). The first was for the VAT photopolymerization technology (US Patent 4575330)[4] and was granted in 1984 to Chuck Hill, founder of 3D Systems, a US-based company. The initial material extrusion technology (US Patent 5121329) was granted in 1989 to Scott Crump, co-founder of Stratasys. Stratasys was a US-based company, but after its merger with Israel-based Objet Ltd. in 2012, it officially based its headquarters in Israel. Stratasys and 3D Systems are among the main companies in the world that manufacture 3D printing (3DP) machines. Third, the initial powder bed fusion technology (US Patent 4863538) was granted in 1989 to Carl Deckard, who was at the University of Texas at Austin. Deckard was one of the founders of DTM. DTM was acquired by 3D Systems in 2001.

As the different technologies show, AM is a generic term for different processes. In 2015, the ISO/ASTM (International Organization for Standardization/American Society for Testing and Materials) 52900 Standard was created to standardize all terminology and to classify each of the different methods of AM. In total, seven process categories were established (Redwood, Schöffer and Garret, 2017) (see Table 2.2). Each of these seven categories can be further divided into sub-categories depending upon a further refinement of the AM process. These seven AM processes are explained in the next sections.

Material extrusion (MEX)

Although chronologically it is not the first AM technology that was developed, the discussion will start with MEX as most people, whether they are conscious of it or not, have familiarity with this process. The discussion will relate to topics that also apply to the other AM processes, such as how AM generally works.

Table 2.2 AM processes

	AM process						
	Material extrusion (MEX)	*VAT photopolymerization (VPP)*	*Powder bed fusion (PBF)*	*Material jetting (MJ)*	*Binder jetting (BJ)*	*Directed energy deposition (DED)*	*Sheet lamination (SHL)*
Materials	Plastics, metal, composite, Paste material	Plastics	Plastics, metal	Plastics, metal, wax	Metal, sand	metal	Paper or composite

Source: author.

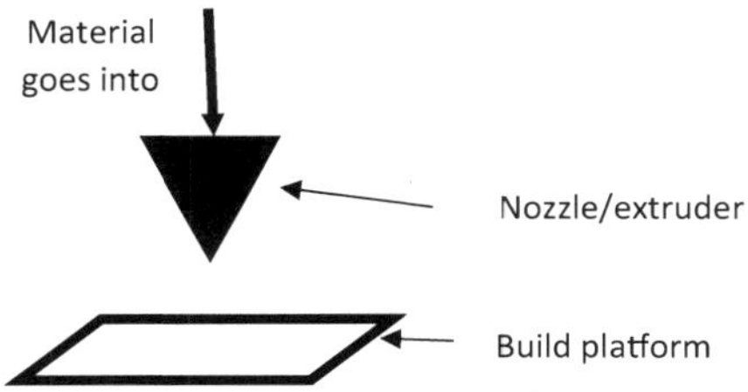

Figure 2.1 Basic schematic depiction of the MEX process.

Source: author.

MEX is similar to pushing toothpaste out of a tube or when a pastry chef puts icing on a cake with a piping technique. In those situations, a material is pushed through a small opening. The pastry chef, by moving the pastry bag and using layers, can create a 3D pattern. MEX is essentially the same process except that it is automated and done by a machine. Figure 2.1 provides a very basic schematic depiction of the MEX process. Material is fed into a nozzle or hot end. The nozzle moves around the build platform (i.e., the area on which the product is manufactured) and, by doing so, builds the product one layer at a time.

Figure 2.2 provides two practical examples of MEX 3D printers. In the pictures, it is easy to identify the build platform and the spools with plastic filament.

The AM process requires (a) material for printing, (b) an AM machine, (c) a digital file of the object that is going to be 3D printed, and (d) software that guides the machine.

a) The material that is 3D printed with MEX can be a paste that does not require heating. For example, a type of batter can be used and then shaped in certain patterns to create a pancake. This is what the PanCakeBot 3D printer does. The use of paste is a type of MEX known as PEM (paste

Figure 2.2 Examples of MEX printers, left the Printrbot Simple Metal, right the Prusa MK3.

Source: author.

extrusion modeling). The material can also be a material that needs to be heated before it can be shaped, such as plastic. This can be used to make things such as figurines. MEX with plastics is known as FFF (fused filament fabrication) or FDM (fused deposition modeling); the latter is trademarked by Stratasys (Lipson and Kurman, 2013; Redwood, Schöffer and Garret, 2017). In the case of plastic, there is a heating element, the hot end, so that the plastic is melted and becomes soft. The soft plastic is 'poured' or formed into the desired pattern by the 3D printer. It then cools off and hardens, thereby leaving the 3D object. Early plastics options for MEX were limited to PLA (polyactic acid) and ABS (acrylonitrile butadiene styrene). ABS is stronger but more prone to warping and thus this required a heated printbed. The printbed is the horizontal surface on which the object is 3D printed. A heated printbed was not included on early desktop 3D printers but since then has become more of a standard feature. There are now many options for materials that can be used with MEX, such as nylon (greater chemical resistance), PETG (polyethylene terephthalate glycol, which is stronger), PEI (polyetherimide, which can handle higher temperatures) and TPU (thermoplastic polyurethane, which is flexible) (Redwood, Schöffer and Garret, 2017). It is also possible to make metal objects with MEX. Companies such as MarkForged and Zortrax are using processes that infuse metal into filament (combination of a plastic with wax) where the plastic and wax are removed in a post-processing sintering process so that metal parts remain.

b) 3D printers can be oriented towards industrial customers (i.e., companies) as well as towards consumers or households. The machines that are targeted at consumers are often called desktop machines or personal 3D printers. Desktop printers are defined as selling for less than $5,000 (Wohlers Associates, 2020). The majority of desktop 3D printers are based on MEX. A consumer who uses AM (thus participates in the production process) and also consumes, thereby blurring the line between consumption and production activities, is called a prosumer (Rayna, Striukova and Darlington, 2015). An example is a consumer or household that buys a 3D printer to manufacture parts or products at home for their own consumption. Industrial AM machines are used primarily by companies, while desktop 3D printers are used by prosumers as well as companies.

   One key characteristic of a 3D printer is the size of the object that can be manufactured. This relates to the build volume, which is the length of the horizontal x-axis and y-axis multiplied by the length of the vertical z-axis. Machines with a small build volume, such as the Monoprice Select MP mini, which has a build volume of 4.7″ × 4.7″ × 4.7″ (or 120 mm × 120 mm × 120 mm), can be purchased for less than $200. However, MEX machines can also be quite large. An example is the Builder Extreme 2000 Pro (with a build volume of 700 mm × 700 mm × 1700 mm), which is priced above $25,000. MEX machines are also used in the construction industry, where their build volume is even larger.

c) In order for a 3D printer to produce an object, it requires that a digital file of the object be available. The majority of these digital files have the STL[5] extension although alternatives such as the OBJ[6] extension exist. There are also developments for other type files (more on this in Chapter 11). There are essentially three ways of getting the digital file. First, the file can be created either by a company that will custom-design for a customer or by doing it yourself. The latter can be done through design software. There are professional design software packages such as AutoCAD and Solidworks. With the growth of prosumer 3DP, (free) internet-based options have also become available. An example is TinkerCAD. This is easy to use, contains a tutorial and can quickly turn anybody into a designer of simple objects. Figure 2.2 provides a simple example of creating a digital file for a hollow cube. This is done by using the standard solid cube shape in TinkerCAD with dimensions of 20 × 20 × 20 mm with another standard but slightly smaller cube shape. The latter cube is turned into a 'hole' so that when it is moved inside the solid larger cube it creates a cavity inside the larger cube. This movement is shown in Figure 2.3; the result is a hollow cube. Thus, this is a very simple process. If some dots were included in the design, one could easily create their own dice!

   The second option for a digital file is to use existing files. For instance, many people design products and make these designs available for others. Sometimes they ask for a payment, but in many instances, the files can be downloaded for free. Examples of such file depositories are

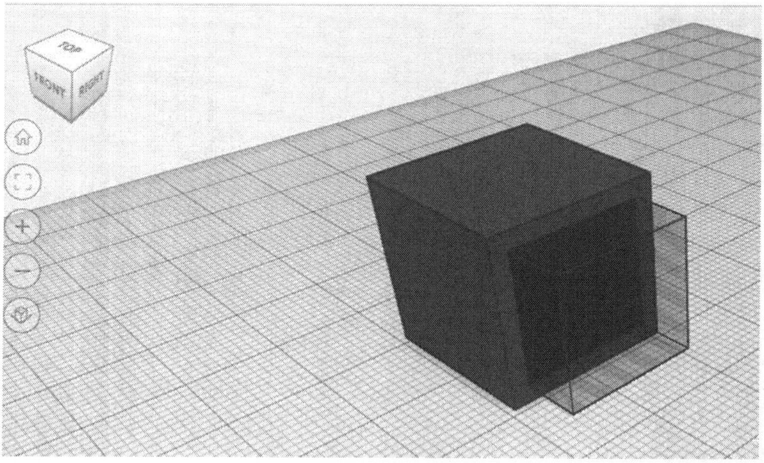

Figure 2.3 Creating a hollow cube in TinkerCAD.
Source: author.

www.youmagine.com and www.thingiverse.com (see Figure 2.4). Youmagine was initiated by Netherlands-based Ultimaker, one of the most well-known companies that produce desktop MEX printers. Thingiverse was initiated by Brooklyn-based Makerbot, one of the early companies that made 3DP available to consumers. It was acquired by Stratasys in 2013. Makerbot and Ultimaker merged in May 2022 and became UltiMaker. In 2019, Czech-based Prusa, which has become quite popular with its desktop 3D printers, created its own 3D design depository (www.prusaprinters.org).

The third option is to create a digital file from an existing object by scanning the object. For instance, the Matter and Form 3D scanner V2 allows objects of approximately 9.8″ tall and 7″ diameter (25 cm high and 18 cm diameter) to be scanned and is available for roughly $750. Thus, after scanning an existing object (let's say a garden gnome or a small vase), the software creates a file that allows these objects to be duplicated with 3DP. Figure 2.5 is an example where an object was scanned with the Shining 3D Einstar scanner and then 3D printed at half the size.

Scanners vary in terms of technological capabilities (i.e., scan accuracy and the size of the objects that can be scanned). Most scanners are aimed at smaller objects, but Arctec3D provides scanning booths that are capable of scanning people. Thus, instead of getting a two-dimensional picture of somebody, an alternative is a 3D printed version. An example of this type of application is scanning a couple that is going to get married, then 3D printing a miniature version of the couple based on this scan, and placing this on top of the wedding cake, giving the wedding cake a personalized touch (Brown, 2019). Online (and sometimes free) applications such as

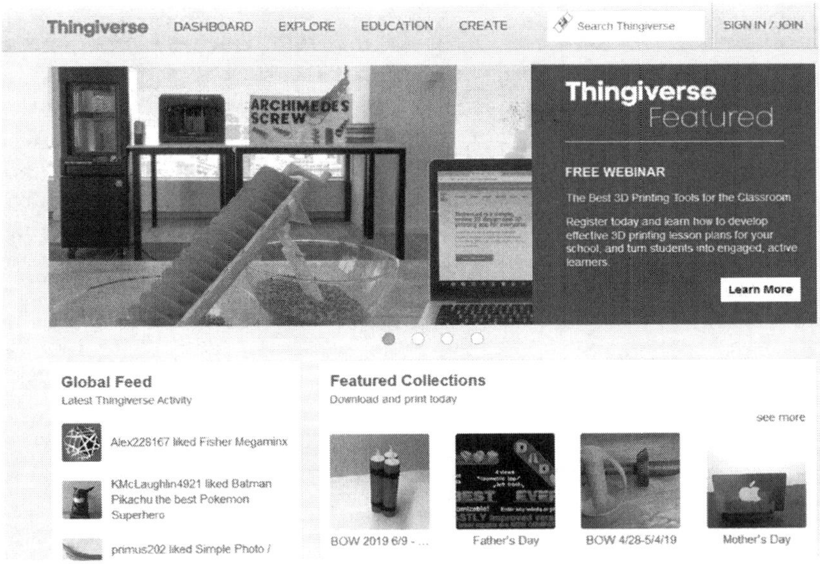

Figure 2.4 Thingiverse opening page.

Source: author.

Figure 2.5 Scanned and then 3D printed object.

Source: author.

Trnio also exist. Scans may require additional processing before they are ready for 3DP. A special case of scanning is that in medical settings because there are already established scanning processes such as magnetic resonance imaging (MRI), computed tomography (CT) scan, and ultrasound. The scans from these processes can be used to create digital files for 3D printing. An example is 3Dprintedultrasounds, a company that provides 3D prints based on pregnancy ultrasounds. So instead of getting an ultrasound picture of the baby in a womb, expectant parents can get a 3D printed version.

d) The software that guides a printer relates to two aspects that need to be handled by the software. The first is the translation of the digital (STL) file of a 3D object into layers. The second is the translation of those layers into movements for the 3DP machine. This is known as the G-code. The G-code is similar to that used in CNC machines and it directs the 3D printer.

Going from a digital 3D file to layers is called a slicing process. Several programs are available to do this, such as Slic3r for desktop 3D printers. Some producers of desktop 3D printers developed their own slicer software: Cura was developed by Ultimaker, and PrusaSlicer was developed by Prusa. These slicers are often integrated in the software of the 3D printer so that once a digital file is uploaded, it can be sliced while a G-code can be created for that particular type of machine.

Some companies use open-source packages for their 3D printers. For instance, the now-defunct Printrbot used Repetier-Host. Purchasers of a Printrbot Simple Metal 3D printer had to download the free Repetier-Host software on their computer. The Printrbot Simple Metal would then also be connected to the computer, and the Repetier-Host software would recognize this type of 3D printer. Once a digital file is uploaded and sliced, it is essentially an issue of 'pushing print' and the 3DP process occurs.

The slicing is of key importance to the entire process of 3DP as it affects the quality and efficiency of printing. Embedded in the Slic3r process are well over 100 variables, such as speed of different types of movement, the temperature of the hot end, and the temperature of the build platform. These are material- and printer-dependent. As the material quality is getting more consistent, 3D printers get more sophisticated, and with an expansion of the types of material available for MEX, there has been a standardization process leading to fewer standardized slicing options. Instead of having to provide input for over 100 variables, the user is limited to a handful of standard options. With this, the software selects the most appropriate settings to use for its slicer and thus for the subsequent 3DP process. Figure 2.6 provides an example of a simple object: a clip designed for bookshelves. The software shows on the top right side that there are only a few options, such as print speed, type of filament, and fill density. Below that is a row of three buttons; the middle one slices the object, and once this is completed, it can be exported (the right-side button) to a file which contains the G-code. This file has to be uploaded to the 3D printer so that the object can be 3D printed.

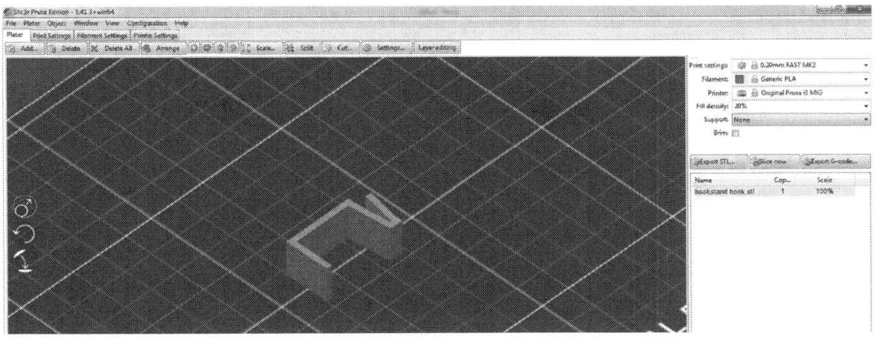

Figure 2.6 Example of slicer for Prusa.

Source: author.

Table 2.3 Material extrusion

	Advantages	*Disadvantages*	*Type of material*
MEX	Relatively simple and can be relatively low-cost. This makes it especially suitable for rapid prototyping where the product is not for end use.	May need supports (more on this later) and post-processing and can be sensitive to warping, and quality depends upon how well the layers stick together.	Increasing assortment of plastics and paste

Source: author.

Some companies, such as Ultimaker, have formed alliances with leading material manufacturers, such as BASF, DuPont, DSM, Owens Corning, and Clariant, which have developed print profiles for their own 3DP materials that are optimized for Ultimaker software and hardware. Thus, the material specifications are already part of the automated process of slicing.

VAT Photopolymerization (VPP)

The general characteristics that were described for MEX, such as related to the files and slicing, also apply to VPP and the other AM processes. However, there are some differences in how the 3DP process happens. For VPP, there is a container that contains a liquid photopolymer. The photopolymer is a chemical that reacts to light. In the 3DP process, light is used and directed at a specific layer in the liquid. In the process, the layer is solidified. This process is then repeated. So, a liquid is made solid rather than, for instance, pouring heated and soft plastic as in MEX. Figure 2.7 provides a very basic schematic depiction of the VPP process. In Figure 2.7, a top-down approach is shown: the light source is depicted above the build platform. The build platform

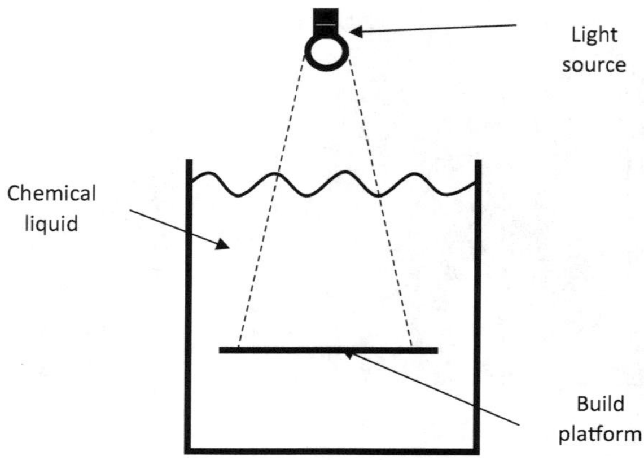

Figure 2.7 Basic schematic depiction of the VPP process.
Source: author.

begins at the top of the container with liquid and moves down layer by layer. The finished product is, therefore, sitting on top of the build platform. With VPP, the light source could also be below the build platform, a bottom-up approach. Here the platform starts at the bottom with just one layer in between the bottom of the container and the build platform. It then cures this layer, and the build platform keeps moving up. The finished product is, therefore, 'hanging' below the build platform.

VPP is used mainly for printing plastics. There are two different VPP methods: SLA (stereo lithography) and DLP (direct light processing). The difference between these two methods is the source of light. For SLA, the source of light is a laser. For DLP, the source of light is a digital light projector screen to flash a single image of each layer all at once rather than tracing it with a laser point. Because of this, DLP can achieve faster print times compared with SLA, but a disadvantage is that owing to the technology of flashing an entire layer at once, the size of the object that can be created is limited (Redwood, Schöffer and Garret, 2017). A specialized form of this process is CLIP (continuous liquid interface production), which is commercialized by a company called Carbon. CLIP works by projecting light into the container but in a continuous fashion; thus, the product is manufactured continuously rather than layer by layer. Carbon has been in the news because of its work with Adidas, which uses Carbon's machines to print midsoles for its athletic shoes. Figure 2.8 shows some of Carbon's machines and also gives an impression of what VPP machines look like. Carbon is very successful and reached the status of unicorn, which is when a privately held start-up hits $1 billion in value. Formlabs, which also uses VPP technology, also reached the status of unicorn.

While generally VPP is used for plastics, there are exceptions to this rule. It can also be used for ceramics. Ceramics are important for medical

Figure 2.8 Carbon's 3D printers.

Source: author.

Table 2.4 VPP

	Advantages	*Disadvantages*	*Type of material*
VPP	Capable of printing fine detail and with smooth finish. Can be used for end products such as hearing aids.	May need supports and post-processing. Quality depends upon material which can be fragile.	Mostly plastics

Source: author.

applications where the advantage over plastics or even titanium is that these latter materials degrade and thus would end up in the body whereas ceramics do not. Ceramics are also used in aerospace and are preferred over metals because there is often heat involved and metals expand and shrink whereas ceramics do not. For this reason, 3D printed ceramic molds are attractive for metal casting processes. It should be noted, however, that this is done in multiple steps, which include that the 3D printed ceramic product is afterwards put in an oven for debinding and sintering (Truxova et al., 2020). Companies such as Lithoz (DLP) and 3DCeram (SLA) make ceramic parts using VPP.

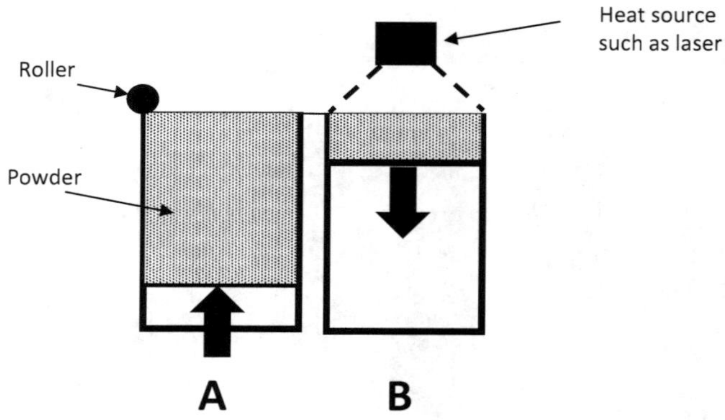

Figure 2.9 Basic schematic depiction of PBF.
Source: author.

Powder bed fusion (PBF)

PBF is an AM process in which powder is used. The powder is fused together in layers and thus the product is created. Figure 2.9 provides a very schematic depiction of the process. The process involves two containers, A and B, with movable bottoms. At the beginning of the print, container A is filled with material in the form of powder, and the bottom of the container is in the lowest position. Container B is empty, and the bottom of container B is in the highest position. When the process starts, the roller moves over one layer of material from container A to container B. The layer of powder is exposed to a heat source such as a laser. Because of the laser, the powder gets melted or fused together and thus the first layer of the product is manufactured. The bottom of container A moves up the height of one layer while the bottom of container B moves down the height of one layer. Then the process repeats itself. Essentially, this means that material is moved layer by layer from container A to container B. When it is moved, it gets exposed to heat so that powder fuses together in the desired shape. At the end of the process, container A is empty and the bottom of container A is in the highest position while container B is full with the product surrounded by powder that was not fused together and the bottom of container B is in the lowest position.

There are currently five recognized types of PBF. Two are used mainly for plastics: SLS (selective laser sintering) and MJF (multi-jet fusion). Three are used for metals: SLM (selective laser melting), DMLS (direct metal laser sintering) and EBM (electron beam melting). The principles behind these processes are the same, although how the process is accomplished is different. SLS and SLM are very similar, but the power of the laser for SLM is much higher because the energy needed to melt and fuse metal powder is much higher than the energy needed to melt and fuse plastics. The difference between DMLS and SLM is that with SLM the metal powder is melted

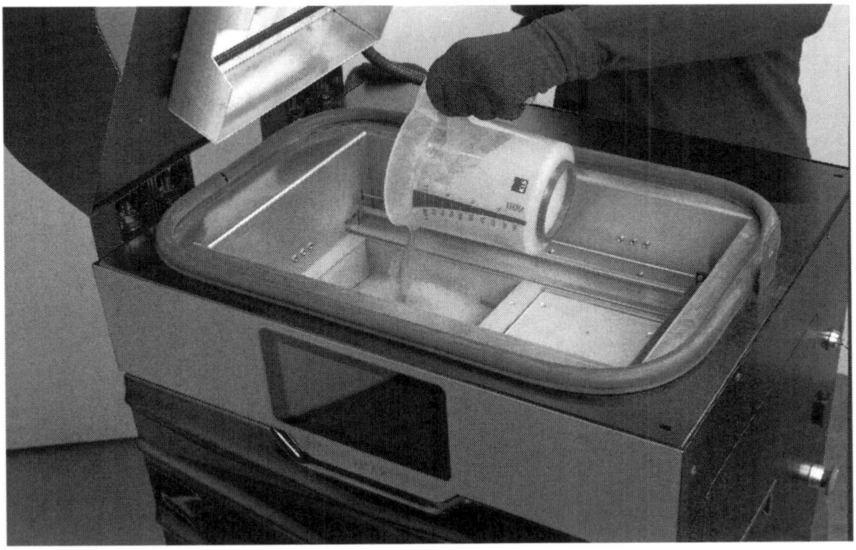

Figure 2.10 Sinterit Lisa Pro.

Source: Sinterit.

whereas with DMLS the powder is heated to a point so that it can fuse together at the molecular level. With EBM, a high-energy beam is used rather than a laser, whereas with MJF, there is an additional selectively applied fusion agent which helps the energy absorption. Figure 2.10 shows an example of an SLS printer from Sinterit: the Lisa Pro. It shows how the left-hand container is being filled with powder. The bottom of the right-hand container is all the way at the top.

For SLS processes, the strength of the laser is important. The Sinterit Lisa has a 5W laser. This can do mostly black materials, but in order to use white materials, which reflect, a stronger laser such as 30W or 60W (e.g., the XYZprinting MfgPro236 xS) is required. Also, in order to use metal (SLM or DMLS), a closed environment without oxygen (but with other inert gases) is required to minimize oxidation of the metal powder. Exposure to oxygen causes embrittlement or other undesirable outcomes, and the more reactive the material (such as titanium), the worse the effect.

Figures 2.11–2.14 provide illustrations of the capabilities of SLS. Figure 2.11 shows a small 3D printed book, 40 mm high, that is made from a nylon-based powder and has clearly separated pages! This demonstrates the technical capabilities of, in this case, the Sinterit SLS process.

Figures 2.12 and 2.13 show examples of a complex shape that is impossible to produce in one piece with TM. Figure 2.12 is printed with a nylon-based material, whereas Figure 2.13 is printed with a flexible material, as is demonstrated in Figure 2.14.

Figure 2.11 Example of a print made with the Sinterit SLS printer.
Source: Wojciech Hajduk/Sinterit.

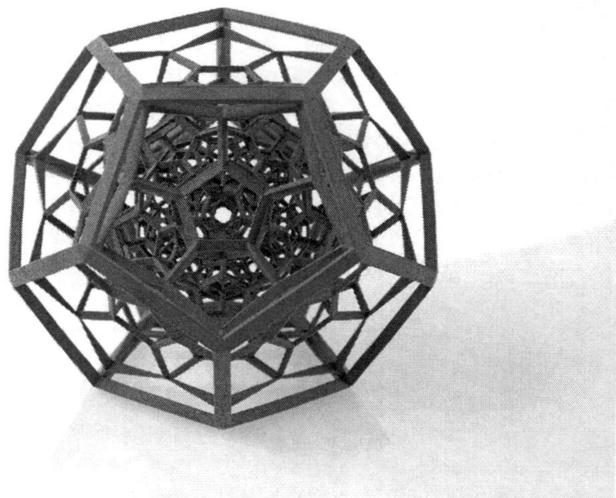

Figure 2.12 Example of an impossible print made with the Sinterit SLS printer.
Source: Wojciech Hajduk/Sinterit.

Figure 2.15 illustrates that the SLS process requires post-processing such as cleaning the remaining powder from the product. One of the advantages of SLS versus MEX is that with SLS there is limited need for support material because the powder that surrounds the 3D printed object provides support.

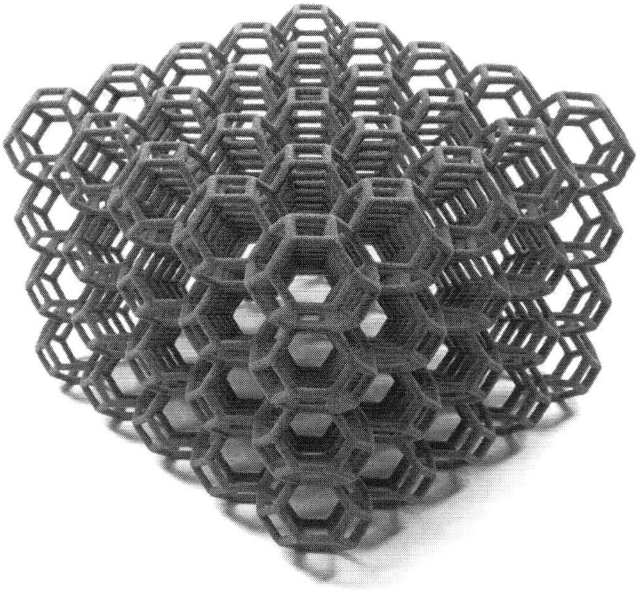

Figure 2.13 Example of an impossible print made with the Sinterit SLS printer and flexible material.

Source: Wojciech Hajduk/Sinterit.

Figure 2.14 Example of flexibility of a print made with the Sinterit SLS printer.

Source: Wojciech Hajduk/Sinterit.

Figure 2.15 Post-processing.

Source: Sinterit.

Table 2.5 Powder bed fusion

	Advantages	*Disadvantages*	*Type of material*
PBF	Capable of making strong end-use parts with complex geometries. Does not require support structure (except SLM/DMLS).	Can be sensitive to warping and requires post-processing. Relatively expensive machines.	Plastics, composites, metals

Source: author.

Material jetting (MJT) and binder jetting (BJT)

The next two AM processes—that is, MJT and BJT—will be discussed together in this section because they have similarities. A key characteristic in both processes is that they involve droplets that are selectively deposited. With MJT, they are tiny droplets of material that are deposited on a build platform (which is similar to inkjet printing but in three dimensions); with BJT, they are droplets of a binding agent (a type of glue) that are deposited in a powder bed to bond the powder. MJT is used for plastics (the plastic is cured with ultraviolet light) but also for wax (wax droplets are deposited and, through cooling, solidify). BJT is used for metals, and as with metal MEX, the bonded material is sintered afterwards. BJT is also used for sand or gypsum without a sintering step. A specific application is using BJT for a sand mold. The sand mold can then be used in a (metal) casting process. A type of MJT process is NPJ (nano particle jetting). This is also used for

Figure 2.16 Basic schematic depiction of MJT.
Source: author.

Figure 2.17 Mimaki 3DUJ-553 MJT printer.
Source: Mimaki.

metals, and as with BJT and metal MEX, the material is cured with heat and sintered afterwards. Figure 2.16 provides a basic schematic depiction of MJT. The difference with BJT is that for the latter the binding droplets are deposited in a container with powder.

Figure 2.17 shows an example of an MJT printer from Mimaki: the 3DUJ-553.

One of the advantages of the MJT process is that owing to working with droplets it is very precise. It is considered the most accurate form of 3DP (Redwood et al., 2017). It is also possible to create color prints by using different colors for different droplets. The Mimaki 3DUJ-553 printer uses water-dissolvable support materials and six types of inks: clear, white, cyan, magenta, yellow and black. With this combination of inks, more than 10 million color combinations can be printed. Figures 2.18 and 2.19 give an impression of what is possible with 3D color printing.

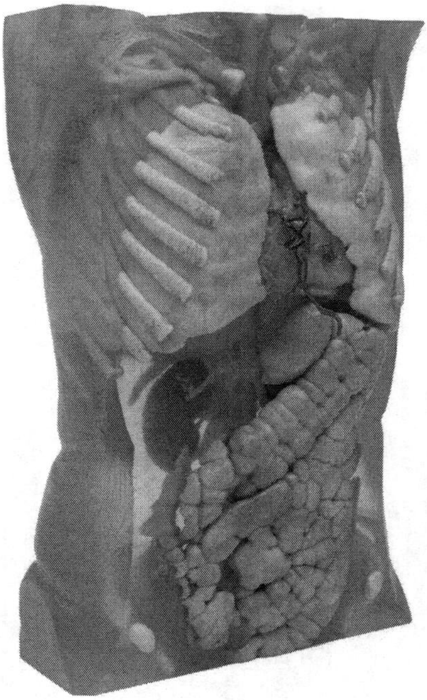

Figure 2.18 color print of a torso with the Mimaki 3DUJ-553 MJT printer.
Source: Mimaki.

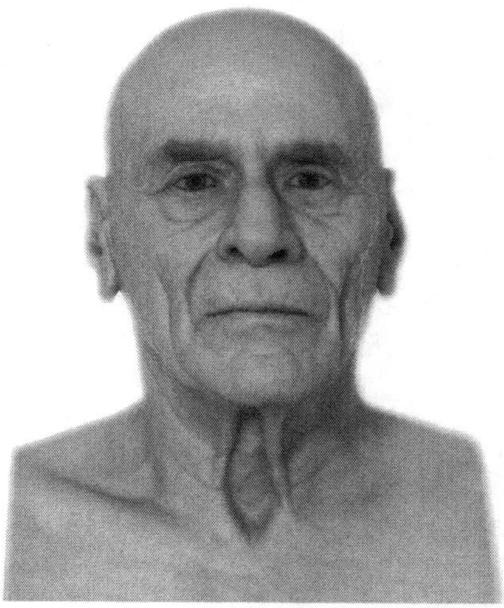

Figure 2.19 3D printed face with Mimaki 3DUJ-553 MJT printer.
Source: Mimaki.

Table 2.6 MJT and BJT

	Advantages	*Disadvantages*	*Type of material*
MJT	Very precise, can print multi-material, colors, and fast. It requires supports, but supports are dissolvable.	Requires materials that can be jetted. Generally limited post-processing. Products can be fragile. Expensive.	Plastics
BJT	Allows complex metal shapes	Requires post-processing such as sintering. Limited strength of parts.	Metal, sand

Source: author.

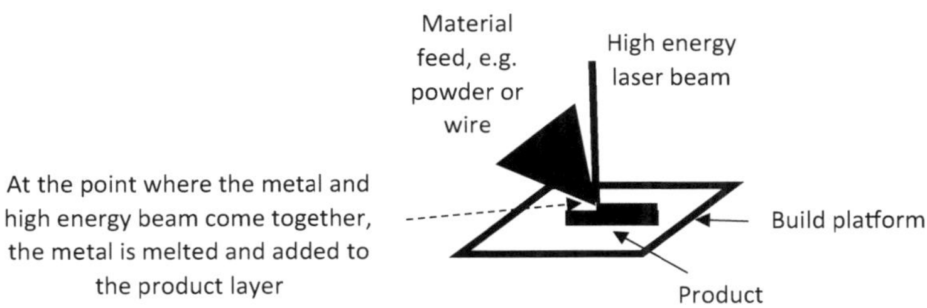

Figure 2.20 Basic schematic depiction of DED.

Source: author.

Directed energy deposition (DED) and sheet lamination (SHL)

The last two types of AM processes—DED and SHL—are not as widespread across different industries. With the DED process, focused thermal energy such as through a laser is used to fuse materials by melting them as they are deposited (see Figure 2.20). DED is used mainly for metals, and the materials can be supplied as powder or a metal wire. There are three main processes using different energy sources: laser engineering net shape (LENS), electron beam AM (EBAM) and wire arc AM (WAAM). LENS is where a metal wire or overblown metal powder is melted by means of an energetic laser beam and selectively applied locally. EBAM is where a metal wire is melted by means of a moveable high-energy electron beam and selectively applied. WAAM also uses a metal wire, but the energy source is an electric arc (similar to welding). An advantage of DED is that the object can be easily moved during the manufacturing process. Owing to this, it is possible to print without supports. Figure 2.21 provides an example.

SHL is a two-step process. In the first step, a sheet of material is put on top of a previous layer and through an adhesive gets bonded (i.e., laminate). In the second step, the contour is cut out using, for instance, a laser or hot wire. This is schematically depicted in Figure 2.22. Figure 2.22 shows two rolls of

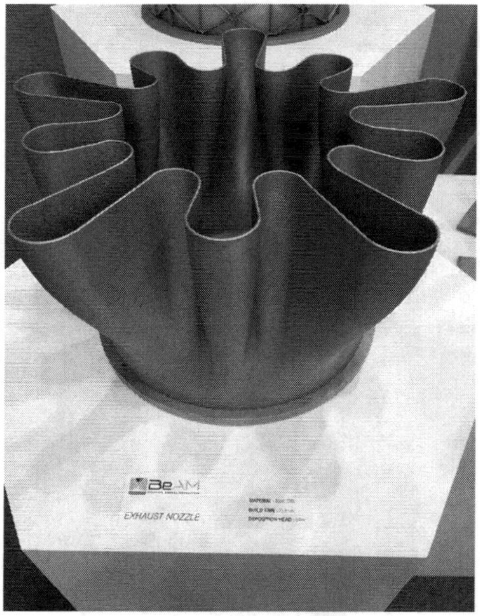

Figure 2.21 Exhaust nozzle produced by BeAM with DED.
Source: author.

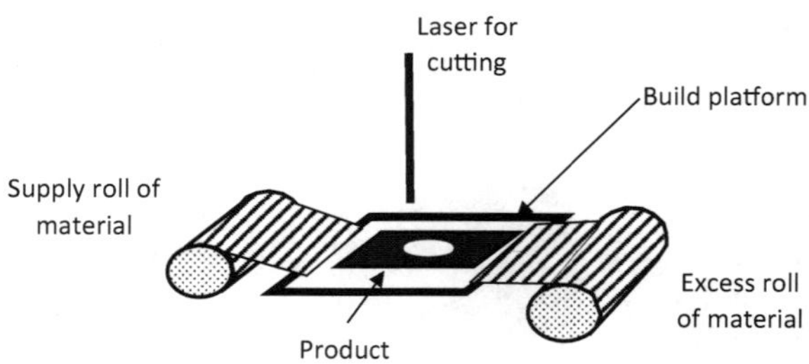

Figure 2.22 Basic schematic depiction of SHL.
Source: author.

Table 2.7 DED and SHL

	Advantages	*Disadvantages*	*Type of material*
DED	Suitable for very large complex shapes, no supports needed. Strong parts.	Low resolution, shielding may be necessary, can be susceptible to distortions, may need post-processing such as polishing or CNC machining.	Metals
SHL	Low cost and fast	Least resolution of all AM types, creates waste, limited in terms of complex shapes	Paper, metal

Source: author.

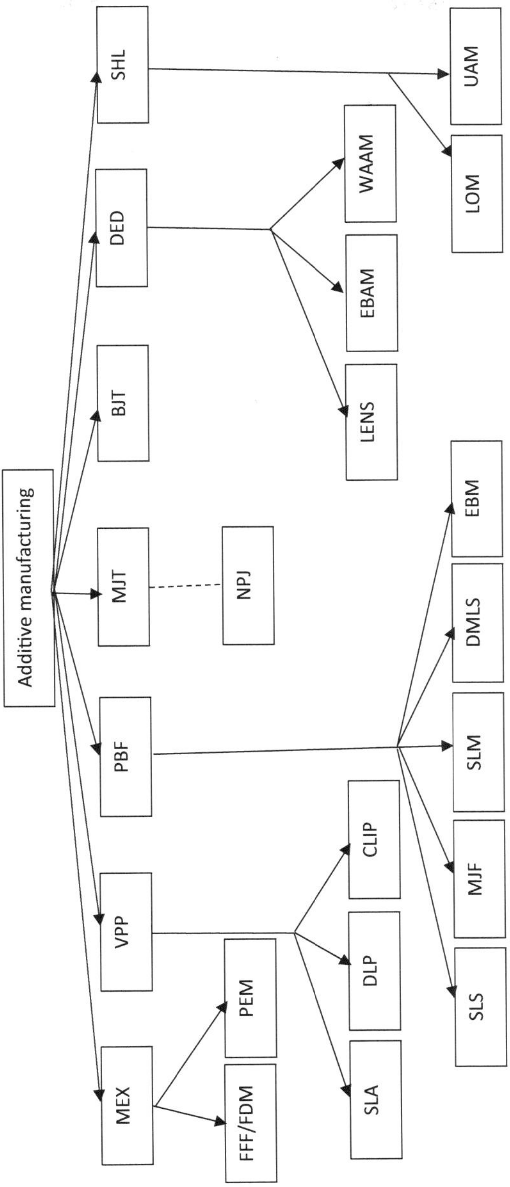

Figure 2.23 Overview of AM types and technologies.

Source: author.

Table 2.8 Overview of different AM types

	Cost	*Quality*	*Time*	*Flexibility*	*Post-processing and waste*	*Main material*	*Main use*
MEX	Low-cost equipment. Low-cost material.	Relatively low resolution. Susceptible to warping. Vertical strength issues.	Relatively fast	Can produce some complex shapes by using bridging and supports	Limited post-processing (e.g., for removing supports)	Plastic	Small volume of parts that do not have stringent quality requirements
VPP	Relatively low-cost equipment Somewhat expensive materials	Smooth surface. High resolution (detailed features). Relatively weak parts.	Relatively fast	Can create some complex shapes but needs supports	Requires post-processing (e.g., for removing supports)	Plastic	Small volume production of parts with detailed features that don't require a lot of strength.
PBF	Equipment a bit more expensive. Material expensive. Post-processing which adds cost.	Strong parts. Susceptible to warping.	Relatively slow	Can do complex shapes, no supports needed	Post-processing required	Plastics and metals	End-use parts.
MJT	Somewhat expensive equipment. Somewhat expensive material.	Very high resolution. Relatively weak parts	Fast	Complex shapes (wax supports that can be melted away). Color. Multi-material	Post-processing such as removal of support structure	Plastic	Small volume production of parts that do not have stringent strength requirements.
BJT	Somewhat expensive equipment and material	Very high resolution. Relatively weak parts	Fast	Complex shapes	Post-processing	Plastic	Sand casting molds
DED	Expensive equipment	Large, strong parts (no supports needed).	Fast	Complex shapes. Can add to existing parts.	Post-processing in particular due to poor resolution	Metal	End-use parts
SHL	Relatively low cost	Low resolution. (No supports needed.)	Fast.	Limited ability for complex parts.		Paper	Models for architectural projects

Source: author.

material as material is moved from the roll on the left, across the object to the roll on the right. With an adhesive, the new layer is connected to the object, and excess material is removed. The laser is then used to cut the new layer into the desired shape. In Figure 2.22, this is depicted by a hole in the object. There are several different types of SHL. Two of the most well known are laminated object manufacturing (LOM), which uses paper and an adhesive, and ultrasonic AM (UAM), which uses sheets of metal which are bound together by ultrasonic welding.

Combination

One of the disadvantages of the processes as explained above is that each has pros and cons and works with different materials. Sometimes manufacturers are producing products that are not based on one type of material or that require multiple types of processes. In AM, this can be achieved by combining the different AM types. For instance, Additive Industries, a Netherlands-based company, develops industrial 3D printers that are based on modules. Customers can configure the system with different modules to fit their applications.

Concluding remarks on AM technology

Figure 2.23 provides an overview of the different AM technologies, and Table 2.8 summarizes the information that was presented in this chapter. The table focuses on key aspects of each of the AM process types to highlight the main differences. There are some exceptions to the general characteristics of these AM processes, such as the use of ceramics in aerospace, which connects to the materials side. However, not every AM process type can be used for ceramics (see also Chapter 11).

Notes

1 But this depends upon the type of AM, as will be shown when the technology is discussed in-depth below.
2 https://3dinsider.com/3d-printing-history/
3 https://www.fabbaloo.com/2018/10/final-word-the-first-3d-printer
4 Interested readers can look up these patents by visiting https://www.uspto.gov/ and doing a patent search by patent number.
5 STL is widely believed to be an abbreviation for STereoLithography, although it is also referred to as Standard Triangle Language or Standard Tessellation Language.
6 "OBJ or "Wavefront Object" is a type of geometry definition file that is named after the computer graphics company Wavefront Technologies.

3 The AM business case I

Comparing AM and TM characteristics

With an understanding of the AM technology from Chapter 2, it is now possible to get initial insights into why AM is attractive for manufacturing companies or, in other words, to get an initial understanding of the business case for AM. This chapter will delve into two aspects: the processes used in manufacturing and how manufacturing companies compete in the marketplace.

Process choice

Manufacturing companies have several strategic choices and decisions to make in terms of how they go about doing things. One of the important decisions relates to what type of production process will be used. Roughly speaking, this has something to do with the variety and volume of the items that will be produced. In order to do so, five generic process types have been distinguished (Hill, 2000) which relate to the (traditional) trade-off between volume and variety (McDermott, Greis and Fischer, 1997). The five generic process types are:

- Project: companies that produce large-scale, one-off (unique), complex products will normally provide these on a project basis. An example is the construction of a new and unique building.
- Jobbing: used to meet one-off order requirements of customers (e.g., tooling made in line with the specific requirements of a customer).
- Batch: for providing similar items on a repeat basis (i.e., larger volumes than jobbing with variety between batches).
- Assembly line: with higher volumes, investment can be made for dedicated lines to produce a single product or a small range of products such as for cars. Semi-complete cars are on a conveyer belt that passes through different stations where additional work is added to the car until, at the end of the assembly line, the car is complete. There is limited flexibility to produce a large variety of cars on an assembly line because there is a dedication to specific work activities that have to be performed on a car.

DOI: 10.4324/9781003399100-4

- Continuous process: a basic material is passed through successive stages or operations and refined or processed into one or more products. Examples are sugar and oil refining. There is very limited variety but a high-output volume. Switching to produce a different type of product with the same process is not realistic.

This is depicted in Figure 3.1, in which the circles indicate roughly where process types are placed. Thus, a project has high variety but in low volumes, whereas a continuous process has high volumes but in low variety. AM influences the traditional volume versus variety trade-off because AM allows high variety even for somewhat higher volumes. That means that AM is more in the top right quadrant of Figure 3.1. Furthermore, for unique products, AM can produce faster results.

Technical capabilities – molds

AM, owing to its additive nature, presents different capabilities than TM processes. This can be explained by looking at a hollow cube. In TM, a plastic cube can be manufactured using a plastic injection molding (IM) machine. This machine works by having liquid plastic being injected into a mold. A mold is something that helps to shape a product. For IM, the mold is

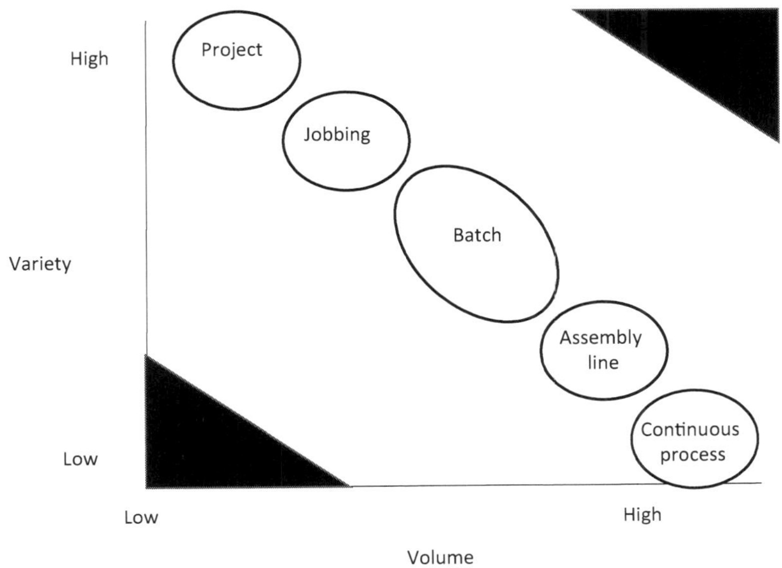

Figure 3.1 Types of manufacturing processes.

Source: Adapted from Hayes and Wheelwright (1984) and Hill (2000).

Figure 3.2 Example of a mold for IM.

typically a large metal block that has hollow areas where the plastic goes. When IM is completed, the mold opens so that the part can be taken out. Figure 3.2 provides an example of a simple mold. On the left side of Figure 3.2, it shows the mold with a completed part in it. On the right side of Figure 3.2, the part is taken out. Figure 3.2 gives a sense of how molds work and the size of molds.

With a mold, it is not possible to create a hollow cube in a single process. Imagine that for the IM machine, the mold has to be used to create the outside walls of the cube. The plastic needs to be formed somehow. The mold can have a cavity that is the shape and size of the outside of the designed cube. But to make it hollow requires a smaller cube (part of the mold) is inside of this bigger cube as part of the mold so that the plastic can be formed around it and so that the plastic cube can be produced that will be hollow. However, such a mold consisting of two parts is impossible because, first, the smaller inside cube would have to be attached somehow to the larger outside cube; without this, the smaller inside cube part of the mold would somehow float in the air, which is impossible. Wherever this attachment would be, it would distort the shape of the plastic cube that is being produced with the IM process. Second, even if it were possible to have a smaller inside cube and to form the plastic product around this, once the plastic product (cube) is formed, it would be impossible to remove the smaller inside cube part of the mold because it would be inside of the plastic product that was formed. It is possible to make a hollow cube with IM but this requires creating two halves of a cube that have to be combined.

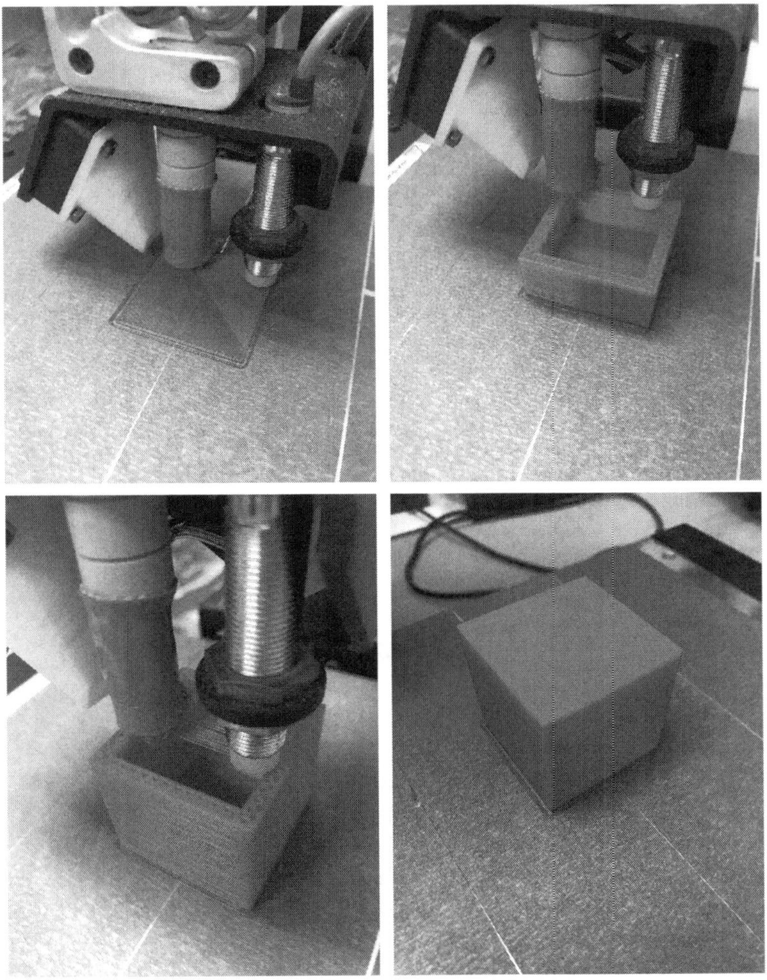

Figure 3.3 Example of making a hollow cube with MEX.

Source: author.

With AM and MEX, it is fairly simple to produce a hollow cube (see Figure 3.3). First, the bottom of the cube is formed. This is simply by extruding materials in the shape of the bottom of the cube as shown in the top left picture. Next, the four side walls are created. This is simple too: the hot end starts to move around the perimeter of the cube, thereby creating a bottom layer for the four walls. This is shown in the top right picture. It then moves a layer up, and the process is repeated until the top of the cube is reached. At this point, the cube has a bottom and four walls and is empty on the inside. Now, the top has to be created. This is a little tricky as somehow the 3D printer has to deposit material crossing the distance from one end of the cube to another end of the cube. Since the cube is hollow, there is nothing underneath the material, so it is easy to imagine that the material will simply drop to the bottom of the cube. This may or may not occur. The process of 3DP across empty space is called bridging, and 3D printers have the capability to do so (as shown Figure 3.3 on the bottom left picture). In other words, 3D printers have the capability to extend material from one side to another side and crossing an area in between that has no support. However, this cannot go on for unlimited distances, which means that the bridging capabilities of a particular 3D printer determine what size of hollow cube can be printed. Current desktop printers can easily bridge 50 mm, which is roughly 2 inches. On the bottom right in Figure 3.3, the completed hollow cube is shown.

When bridging is discussed, the issue of supports needs to be understood. In AM, it is sometimes necessary to print with support material. For instance, if the area that needs to be bridged is beyond the technical capabilities of the 3D printer, then support material can be used so that the 'bridge' can still be made. Support material is not part of the object that is printed. After the 3DP process is finished, it is removed in the post-processing stage. Slicers such as the Slic3R for desktop 3D printers provide options to include support material, and when used, the slicer software will determine where support material should be added. Support material is printed thinner and lighter so that it can easily be removed. The advantage of using support material is that it can benefit the quality of the print or even make it possible to print an object that otherwise could not be printed. The main disadvantages are that it adds material, which has to be removed, and that it creates waste material. Supports make the product more expensive and add labor time for its removal.

Supports also relate to overhang capabilities. A MEX overhang is shown in Figure 3.4. This is an overhang test; that is, it can show the capabilities of the 3D printer in terms of what type of overhang or angle can be printed without supports. The upper left picture shows the print without supports. In this example, the technical capabilities of the 3D printer are quite good and there is little drooping in the product. The bottom left picture shows the 3D print in process but now with support materials. What this clearly shows is how the support material is thin and has gaps in between. The upper right picture shows the completed 3D print with support material underneath the product. The bottom right picture shows how the support material is easily removed in the post-processing stage. Some printers have multiple extruders or multiple material capabilities. In that case, it is possible to print support material that dissolves in water. This makes the post-processing stage of removing the support material even easier.

Figure 3.5 shows a comparison of 3D printed overhangs. The product on the bottom was printed on a Prusa MK3 with support material and has a very smooth surface. The product in the middle was also printed on the Prusa MK3 but without support material. There are very few quality issues here

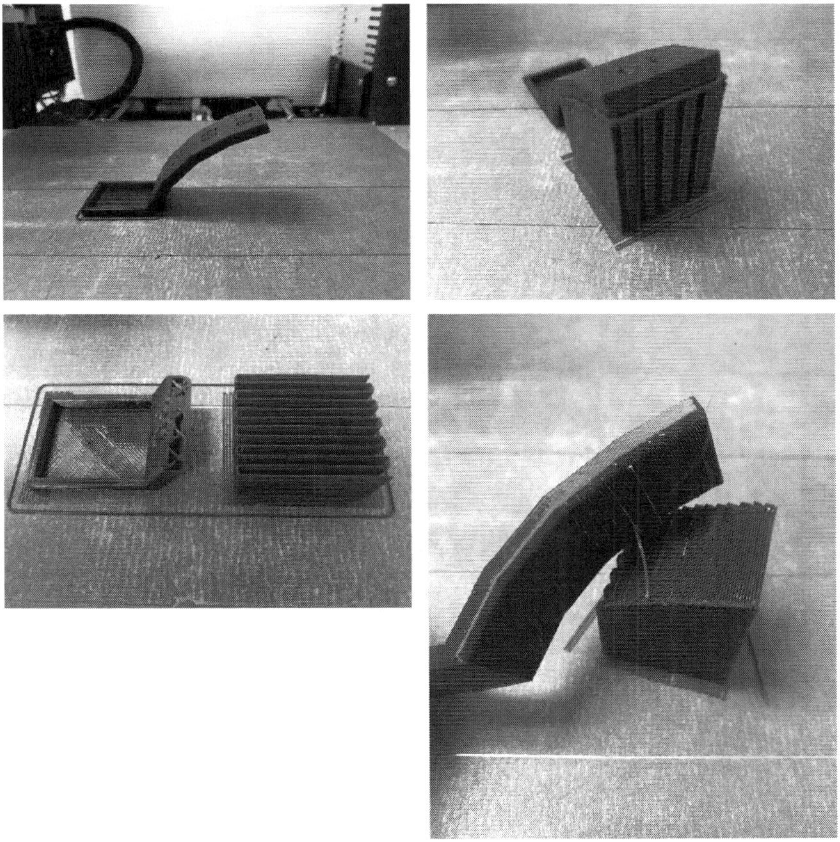

Figure 3.4 Example of printing with support material.

Source: author.

and this demonstrates the Prusa MK3's technical capabilities in terms of printing overhangs. The product on the top was printed on a different printer and has more deformities. This illustrates how different printers can have different technological capabilities.

Capabilities and performance objectives

Slack and Lewis (2011) consider that companies have to strategically reconcile their operations strategy decisions with the performance objectives that reflect the requirements in the marketplace. The operations strategy decisions relate to operations areas such as capacity, supply networks, process technology, and development and organization. Performance objectives relate to

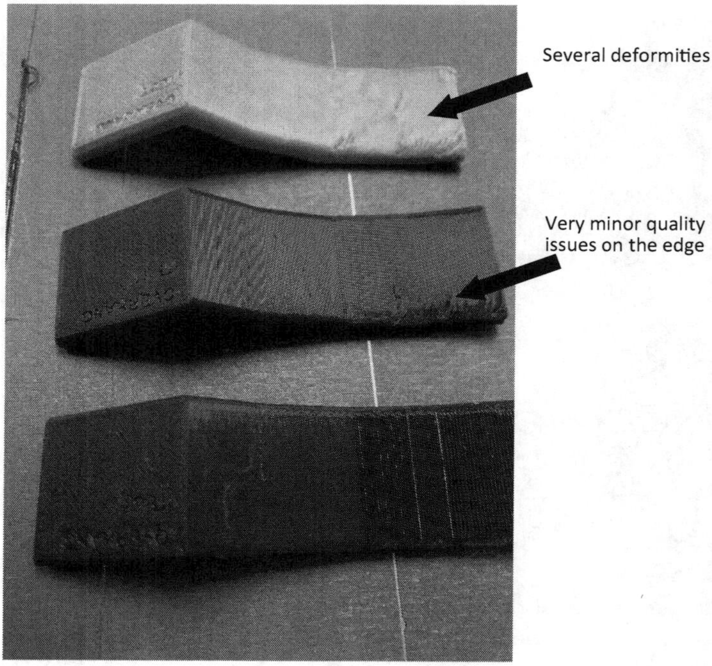

Figure 3.5 Overhang printing with and without support material.
Source: author.

cost, quality, speed, dependability, and flexibility. For example, manufacturing companies make decisions on their capacity, such as the machines they use for production. This affects how they perform in the marketplace. Different types of machines will lead to a different performance such as faster or with better-quality products.

Whereas Slack and Lewis (2011) identify five performance objectives, other studies sometimes recognize fewer or more performance objectives and use different but related terms such as competitive priories and competitive capabilities. Sansone, Hilletofth and Eriksson (2020) have a slightly refined view compared with Slack and Lewis (2011). Sansone, Hilletofth and Eriksson (2020) distinguish six dimensions for competitive manufacturing: cost, quality, time, flexibility, innovation and sustainability. These six dimensions account for 21 operations capabilities (see Table 3.1). Ten of the capabilities that are relevant for AM compared with TM are highlighted in black in Table 3.1. The other capabilities are important for companies but do not specifically address differences between TM and AM. The next sections will explain in general how these ten capabilities are influenced by AM, and illustrations will be provided in part II of this book.

Table 3.1 Operations capabilities influenced by AM compared with traditional methods

Strategic dimension	*Manufacturing capability*	*Definition, that is the ability to*
Cost	Cost efficiency	Provide products at low cost
	Resource efficiency	Maximize the utilization of process resources (machinery and human)
Quality	Product quality	Provide durable, high-performance products
	Process quality	Provide products/services with consistent quality
	Delivery dependability	Deliver on time
	Brand quality	Build a strong and positive company image
Time	Delivery time	Deliver in a short time frame
	Time to market	Have a short time frame from product development to market introduction
Flexibility	Product flexibility	Customize products based on customer requirements
	Product line flexibility	Provide a range of products with different features
	Volume flexibility	Respond to changes in market demand
	Production mix flexibility	Change the product mix in manufacturing
	Labor flexibility	Employees to perform different types of tasks
	Delivery flexibility	Change delivery times and quantities within the agreed-upon delivery time
Innovation	Product innovation	Develop and introduce new products
	Service innovation	Develop and introduce new services
	Process innovation	Develop and implement new processes
	Technology innovation	Develop and implement new technologies
	Market innovation	Find and exploit new markets and opportunities
Sustainability	Product sustainability	Provide sustainable products
	Process sustainability	Manufacture products in a sustainable manner

Source: adapted from Sansone, Hilletofth and Eriksson (2020).

Cost

A key aspect of AM is the different cost structure compared with TM methods which can lead to lower production cost, especially at lower production volumes (cost efficiency). In TM, the cost of manufacturing products is dependent upon the volume of production. This is because specialized equipment and tooling such as molds are used. Molds are relatively expensive and take time to make. Steenhuis and Pretorius (2017) indicate that the price for molds to make items that are just a few inches in size varied from \$6,000 to \$25,000. The cost

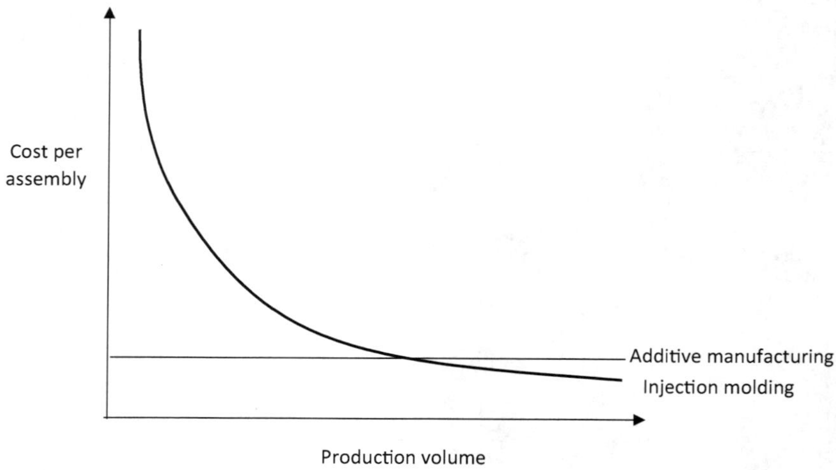

Figure 3.6 Comparison of IM and AM.

Source: adapted from Atzeni et al. (2010).

for these molds has to be spread over the number of items that are produced. A consequence of this is that for TM, the higher the production volume, the lower the average cost per item. AM has a lower need for tools such as molds, so that the cost is less dependent on the production volume. If the same product is produced again with an AM process, then the cost is essentially the same because there is no cost of a tool or mold that has to get amortized over the production run. This concept is illustrated in Figure 3.6, which shows how the cost per part for IM goes down with a higher production volume while the cost per part with AM is constant. Figure 3.6 illustrates that AM is better for lower production volumes while IM is better for larger production volumes.

The height at which the AM line is drawn in Figure 3.6 depends on the price of the AM machine because this influences the cost of making parts. As with molds, the cost of a machine needs to be spread over the production that takes place over the (economic) life of the machine. Let's say there is a \$100,000 machine that is assumed to have an economic life of five years. This means that there is a depreciation of \$20,000 per year. There are roughly 8,760 hours per year (365 days × 24 hours per day), and with a utilization of around 45%, the machine would be used 4,000 hours per year (two eight-hour shifts for five days a week for 50 weeks). The cost per machine hour is thus \$20,000/4,000 = \$5 per hour.

> (Note that with the cost per machine hour information a machine cost per part can be calculated for a set of products which is important when comparing AM to TM. Let's say that one batch of parts take ten hours of machine time and that batch contains 100 parts, then this means that the cost per part is 10 hours × \$5/hour / 100 parts = \$0.50 per part.)

Desktop 3D printers are relatively cheap: the average selling price was $1,196 in 2019 (Wohlers Associates, 2020). They are generally relatively small (can often be carried) whereas industrial systems can be rather large. The Additive Industries MetalFab 1 is several meters long. Desktop 3D printers are based mostly on MEX, although some are based on other AM types such as VPP while industrial systems use more complex AM technologies such as MJT, BJT and DED. The average selling price of industrial systems was $98,105 in 2019 (Wohlers Associates, 2020). However, there is a large variety in the prices of industrial machines, depending upon capabilities and the type of AM. Mimaki's 3DUJ-553 (MJT) list price is $190,000, while some of the Stratasys machines based on MJT sell for around $90,000 and the Xjet Carmel 1400M (MJT) is closer to $650,000. While each of these machines uses MJT, they have very different capabilities. Similarly, industrial PBF machines are generally in the neighborhood of $500,000, although smaller industrial systems are closer to $250,000, whereas machines like the EP-M650 are closer to $1,5000,000. With those prices, the amortization of the cost becomes important and with it the degree of utilization and efficient use of the machine (Basak et al., 2022). A manufacturer needs to have sufficient volume of demand to be able to justify the investment. With TM due to tooling, this means large batches (volume) of the same item, while with AM it can be a combination of (very small) batches with a variety of products.

Table 3.2 provides a cost comparison of IM with AM based on Atzeni et al. (2010). Two different types of AM machines, both based on PBF, were used. Owing to different machine sizes, there is a difference in price of the machine, which affects the production cost. In the study by Atzeni et al. (2010) (and related to Figure 3.6), the point where the cost with IM is the

Table 3.2 Illustration of cost differences between IM and AM (cost in Euros)

	IM with production volume				*3DP with machine*	
	5,000	*20,000*	*100,000*	*500,000*	*Smaller*	*Bigger*
Mold cost per part	**6.240** (97.7%)	**2.900** (96.8%)	**1.075** (93.5%)	**0.335** (84.6%)		
Assembly cost	0.035 (0.5%)	0.035 (1.2%)	0.035 (3.0%)	0.035 (8.8%)	0.012 (0.9%)	0.012 (1.0%)
Machine cost per part	0.083 (1.3%)	0.042 (1.4%)	0.024 (2.1%)	0.014 (3.4%)	**0.694 (58.7%)**	**0.867 (65.9%)**
Material cost per part	0.011 (0.2%)	0.011 (0.4%)	0.011 (0.9%)	0.011 (2.7%)	0.360 (30.4%)	0.382 (29.1%)
Machine operator cost per part	0.018 (0.3%)	0.009 (0.3%)	0.004 (0.4%)	0.001 (0.4%)	0.117 (9.9%)	0.054 (4.1%)
Total cost of product	6.387	2.996	1.149	0.395	1.183	1.315

Source: Based on Atzeni et al. (2010).

same as for AM (where the two lines cross) is 73,000 parts for the bigger machine and 87,000 parts for the smaller machine. Note how the machine cost per part is high for AM and the majority of the cost but that for IM the mold cost per part is high and is by far the majority of the cost (see bolded figures in Table 3.2). It can be expected that the cost for AM machines comes down when more are sold, which will lower the machine cost per part and make AM more competitive at higher volumes.

AM can also lead to lower logistics cost because AM can be applied on the location where it is needed, thus saving on shipping cost. Another way in which AM is less costly than TM is that the former allows for the production of complex items in fewer steps. This reduces the need for assembly and typically the number of parts (and weight), which reduces the cost. AM is generally cost-competitive for low-volume production or for complex parts, although this depends on the type of material and the type of AM that is used. For instance, for metal parts using PBF, Atzeni and Salmi (2012) showed that AM was only cost-competitive below a production volume of 50 units while for plastics AM was cost-competitive for a volume of 87,000 units, as was discussed above.

Another aspect that influences cost is productivity (*resource efficiency*). This relates to issues such as capacity utilization which can be improved by using techniques such as just-in-time production (Sansone, Hilletofth and Eriksson, 2017). The issue of utilization of AM machines is not significantly different from TM, as was explained above. However, one particular aspect can influence the utilization of resources: safety. In general, when workers are satisfied with the working conditions and feel safe from injuries, they become more productive (Karanikas et al., 2018). For that reason, resource efficiency is mentioned here as potentially being impacted by AM in the context of safety.

Quality

A key quality capability is the *product quality* in terms of its performance. AM can provide better-performing products. There are two reasons for this. First, in many instances, the product design for TM methods isn't ideal for its intended purpose but is influenced by manufacturing process limitations. AM allows the manufacturing of complex shapes which can be closer to the ideal (Karnal and Rizza, 2019). Second, sometimes the materials used for the production of a product with TM processes aren't ideal either but, again, are driven by manufacturing process limitations. With AM, increasingly more materials are available which can be closer to the ideal material for a product. An example comes from Verve Cycling, which developed a cycling power meter crankset. With TM, the crankset can be designed only for areas that can be accessed (the outside), but AM has allowed exactly what is needed to be designed. Furthermore, a lighter and stiffer component could be made through the use of a specific type of titanium that is suitable for AM.[1]

While not highlighted in Table 3.1, AM has a disadvantage for *process quality*. Process quality is defined as the ability to provide products/services with consistent quality and conform to specifications. Owing to the relatively young age of AM compared with several TM processes, the consistency of the quality is sometimes problematic (although this continues to improve). The inconsistencies can be small in nature and not visible with the naked eye but can be still be important (Wittwer and Seita, 2022). For instance, when MEX is used, the product that is created is not always exactly the same (Wang et al., 2020; Ye et al., 2021). This is due to minor variations in temperature and in the flow of material and so on. This also means that sometimes errors can occur which result in products that do not conform to the specifications. There are also concerns about the bonding of different layers and whether products created with AM have the same functionality as traditionally manufactured products. Research continues to be conducted in this area so that consistency in product quality continues to improve (see Kuznetsov et al., 2019, Kuznetsov et al., 2020; Santana et al., 2017).

The capabilities of *delivery dependability* and *brand quality* are not specifically affected by AM compared with TM.

Time

AM can improve the ability to deliver in a short time frame (*delivery time*). One the one hand, AM offers the opportunity to produce closer to the market. Thus, raw material can be shipped to a location and then locally products can be made with AM rather than shipping complete products. This can save time.

In terms of production time, AM is slow compared with TM processes. Many metal parts can quickly be shaped by using a press. In such a process, a sheet of metal is put in the press, and by using tools and dies, it can be bent. A search on youtube.com for 'sheet metal stamping' provides videos that give insight into the speed of this process. Metal 3DP through, for instance, PBF is much slower. It involves making an object in layers, which takes more time than doing it all at once such as with a press. Furthermore, each layer takes time as the laser needs to move around. A search on youtube.com for 'metal laser sintering' can give an impression of this process. One way that this disadvantage of AM may be overcome is by using parallel processes in so-called 3DP farms. This is a facility that exists of multiple 3D printers that work in parallel, which reduces the average production time per part. If 3DP of one object takes 30 minutes, then by using 10 parallel printers, 10 objects can be produced in 30 minutes, making the average time 3 minutes. Prusa is an example of a company that utilizes a 3DP farm; it uses over 500 printers.[2]

AM provides a significant speed advantage in prototyping and for creating unique products (*time to market*). With AM, molds are not required, thus saving the time for development, production and shipping of the mold. Prototypes and unique products thus can be created much faster and

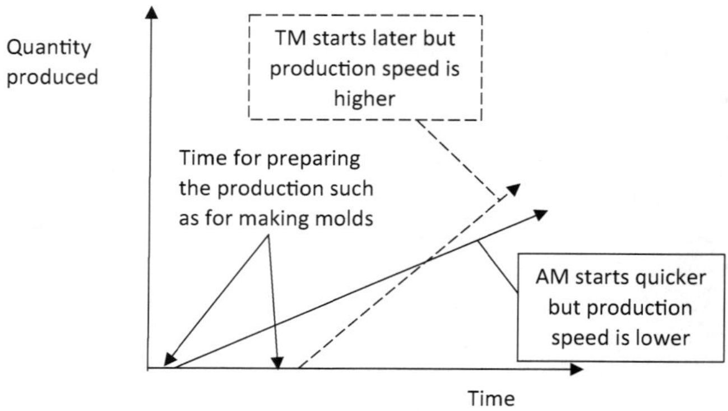

Figure 3.7 Comparing time for AM and TM.

Source: author.

therefore it is no surprise that AM adoption was initially for rapid prototyping (see Chapter 4). For higher production volumes, this advantage of AM over TM diminishes because while it takes longer to start production with TM because of the time it takes for the molds, once production starts TM makes up for this because the TM processes are faster, as explained above. This is illustrated in Figure 3.7. Where the lines cross or whether they cross depends upon the number and complexity of tools such as molds that are required for TM (this influences the preparation time) and the number of items that are produced.

Fast prototyping also means that prototypes can be evaluated quicker. This, in combination with digital files that are easy and fast to modify, means that modifications to the prototype can be made faster so that a better final product can be brought to the market quicker. An example is the quick development and production of mannequins with AM.[3] Even if TM is used for production of the final products, then the time to market can still be drastically reduced by using AM instead of TM for the production of tools and molds. Therefore, AM allows companies to compete more on time (see Stalk and Hout, 1990).

Flexibility

Product flexibility is the ability to customize products on the basis of customer requirements. This is a main direct benefit of AM (Chua and Leong, 2017). The AM technology offers great opportunities to customize products (and at low cost) and has the ability to create complex shapes that cannot be manufactured with traditional methods.

Product line flexibility, *volume flexibility*, *production mix flexibility*, *labor flexibility* and *delivery flexibility* are not highlighted in Table 3.1 since these capabilities are not specifically affected by AM compared with TM. While, for example, AM provides a manufacturer with the ability to provide a wide range of products with different features and to easily have changes in features or product characteristics, the same applies to general purpose machines that are used in TM.

Innovation

Connected to the aspect of rapid prototyping, AM offers the ability to quickly develop new products (*product innovation*), and new services can be developed as well (*service innovation*). An example of a new product is the biopen, a hand-held device that can be used during surgery (O'Connell et al., 2016). An example of a new service is the ability to offer custom-made foot orthoses which increase the well-being of standing workers (Tarrade et al., 2019). For some (i.e., complex) products, AM may be the only viable manufacturing method. Note, however, that this requires design skills. Feedback from the market can be used or ideas can be generated for a new product or service that then can be produced either as a prototype or as a final product. This aspect has played a role with desktop 3D printers where prosumers have come up with all kinds of new products (see below).

Process innovation, *technology innovation* and *market innovation* are not highlighted in Table 3.1. AM is itself an innovation with many consequences, and while these three capabilities relate to the invention of AM, using AM does not affect a company's capability for process innovation, technology innovation or market innovation differently from other types of manufacturing processes or equipment. Note, for example, that utilizing AM can lead to new products that lead to opportunities in new markets. However, companies that use TM can also develop new products, produced with TM methods, that can lead to opportunities in new markets. The difference is the nature of the new market (which falls under product innovation).

Sustainability

TM processes typically generate quite a bit of waste, which is harmful to the environment and is costly. For instance, when a CNC machine is used to cut away material from a block of metal, it leads to waste. AM provides the ability to manufacture products in a (more) sustainable manner (*process sustainability*). Since AM is additive in nature, there is less waste. This is more sustainable and better for the environment and reduces the cost of manufacturing compared with more TM methods. However, note that owing to, for example, the use of support structures as discussed earlier, AM is not free of waste. Related to this is the incorporation of recycling. Pepsi provides an example by considering to recycle bottles into filament for AM. Continued development of materials for AM applications also means that AM offers better *product sustainability* than TM. Some AM processes can use biodegradable materials, while these new types of materials cannot be used for TM

Table 3.3 AM compared with TM for important operations capabilities

Strategic dimension	*Manufacturing capability*	*AM advantage over TM*
Cost	Cost efficiency	AM has more fixed cost structure and has lower need for tools and molds which makes it cost-efficient for lower production volumes. It can also be used closer to the customer saving on logistics cost.
	Resource efficiency	Improvements in safety can lead to higher worker satisfaction and therefore higher productivity.
Quality	Product quality	AM allows product design and materials that are closer to the ideal.
Time	Delivery time	AM has a lower need for tools and molds, thus saving time, which is especially relevant for lower production volumes.
	Time to market	AM has a lower need for tools and molds and therefore a prototype can be quickly produced. Owing to digital files, it is also easier to improve the prototype. This leads to a faster time to market. Even for a product that uses TM, the molds used can be produced quicker with AM than TM, so that this still leads to a quicker time to market.
Flexibility	Product flexibility	AM allows for more flexibility in terms of product design (i.e., customization).
Innovation	Product innovation Service innovation	Owing to the ability of AM for rapid prototyping and quickly making changes to a design, AM allows for quickly developing and introducing new products and services.
Sustainability	Product sustainability Process sustainability	AM generates less waste and can be more economically feasible than TM and uses a wider range of sustainable materials.

Source: author.

processes. An example of this is 3D printed packaging with biodegradable materials (Zeidler et al., 2018; see also Chapter 12).

Table 3.3 provides a summary of the comparison of AM and TM differences.

The case for desktop 3D printers

3DP by prosumers has not developed as much as has been anticipated. Although there are some concerns about the validity of the Gartner hype-cycle (Dedehayir and Steinert, 2016), it can illustrate why expectations were too high (Shi and Herniman, 2023). The hype-cycle shows that after an initial innovation trigger, early publicity produces a number of success stories that lead to inflated expectations. After this, interest wanes as experiments and implementations fail to deliver. Producers of the technology are able to continue to gain investments only if the surviving providers improve their products to the satisfaction of the early adopters. Part of the adoption has to do with technology aspects, uncertainties, and user friendliness. (Chapter 4 will delve more into technology adoption.)

Despite this lower-than-anticipated adoption, 3DP for households makes a lot of sense from an economic perspective (cost efficiency). This has to do with the purchase of a machine (i.e., a capital budgeting decision). Two important criteria that are generally considered are the payback period (PBP) and return on investment (ROI).

> The PBP is the initial payment divided by the expected net income per year for the machine. If an AM machine costs \$100,000 and the expected net income for this machine is \$20,000 per year, then the payback period is \$100,000/\$20,000 = 5 years. In other words, it takes 5 years to earn back the money that was spent on the machine.
>
> The ROI gives insight into the profits that will be earned by the machine and is calculated as the net profit divided by the total investment. If a machine is bought for \$100,000 and the net profit related to that machine is \$10,000, then the ROI is \$10,000/\$100,000 = 10%. Both the PBP and ROI are useful when comparing purchase options for equipment as well as comparing to outsourcing.

A study by Wittbrodt et al. (2013) looked at the decision for a household to purchase a 3D printer. They used common household items such as an iPhone case, garlic press and shower head and then looked at 3D printing them instead of buying. They found that with several conservative assumptions such as printing just 20 items per year, the PBP is less than one year and the ROI was between 20% and 40% (but this depends upon the cost of the 3D printer used).

As an illustration, we lost the coin holder that sits in the middle console of a Saab car. At the time of this writing, this item cost \$12.82 to replace.[4] Instead an alternative, which, though shaped differently, offered the same functionality, was designed (see Figure 3.8). The cost to make it with a desktop 3D printer was roughly \$1.25; therefore, roughly \$11.50 was saved with just this one print.

Desktop 3D printers are also increasingly used in education in many different fields like pharmacy, mathematics, biology, chemistry, art education, graphic design, engineering and even in early childhood and special education (Üçgül and Altıok, 2023). Arvanitidi et al. (2019) identified six reasons for using 3DP in education: to teach students about 3DP, to teach educators about 3DP, as a support technology during teaching, to produce artifacts that

Figure 3.8 Redesigned Saab coin holder.
Source: author.

aid learning, to create assistive technologies, and to support outreach activities. At the primary school level (Pearson and Dubé, 2022) and secondary school level (Ford and Minshall, 2019), AM has been introduced mostly as a way to engage students with STEM (science, technology, engineering, and mathematics) learning. At universities, 3DP has also been used to create test models for experiments (Ford and Minshall, 2019). The identified benefits for using AM in education include greater student involvement, the ability to visualize and better understand theoretical concepts, and the integration of practical and theoretical skills (Assante et al., 2020). While the emphasis in education is mostly on the learning aspects, an advantage of using AM is that it offers flexibility (*product flexibility*) for producing a variety of objects that are economically feasible (*cost efficiency*). Figure 3.9 shows a 3D printed object that can help students to visualize and understand jet engines. This was 3D printed from files made available by user Catiav5ftw on Thingiverse.[5]

Desktop 3D printers have also facilitated creative and entrepreneurial activities (Rayna and Striukova, 2021). This can be seen by the sheer number of files that are available on Thingiverse, YouTube videos of somebody designing and 3D printing his own shoes,[6] and the maker and do-it-yourself movement. The maker movement broadly relates to people making things.

Figure 3.9 3D printed jet engine.
Source: author.

AM has facilitated this because it has allowed smaller-scale and less expensive machines so that prosumers can now make things that, because they required expensive industrial machines, were not previously attainable. This small-scale manufacturing is viewed as a potential tool in economic development (Wolf-Powers et al., 2017). In many instances, people are just sharing their designs, but it can also lead to new businesses. Anderson (2012) describes how 3DP of a drone led to the establishment of the 3D Robotics company. There are many such start-ups facilitated by the application of AM or the development of new AM, as will be shown in part II of this book. It is possible because desktop AM provides an opportunity to experiment with new ideas and designs (*product flexibility*), in an economical manner (*cost efficiency*) and with relatively short time frames (*delivery time and time to market*). This is one of the characteristics of Industry 4.0, namely more dispersed manufacturing. However, on a societal scale, this requires that people have AM skills. Aside from education, there are initiatives such as a European Union–funded project to improve workforce skills in AM.[7]

Notes

1 https://3dprintingindustry.com/news/worlds-first-3d-printed-titanium-cycling-power-meter-crankset-hits-the-road-204735/
2 https://all3dp.com/2/3d-printing-farm-great-showcases/
3 https://3dprint.com/288908/business-development-the-worst-idea-in-3d-printing-ever/

4 https://www.esaabparts.com/saab/parts/4741815
5 https://www.thingiverse.com/thing:1327093
6 https://www.youtube.com/watch?v=dGDzf6ZwwI4
7 https://www.3dprint-training.com/

4 The spread of AM technology

With an understanding of AM capabilities and advantages from the previous chapter, this chapter delves deeper into how a technology gets adopted. Examples such as Betamax, with its superior picture quality over VHS, show that the best technologies are not always successful in the marketplace. The purpose of this chapter is to gain insight into the adoption of technology which gives an indication of the diffusion of AM and what more can be expected. The discussion starts with explaining the importance of innovation and then delves into adoption theory and challenges with high technology.

Technology and the economy

Worldwide economic development was relatively stagnant for thousands of years. The first meaningful sustained economic growth occurred from 1500 to 1700 for The Netherlands, even though the annual growth of 0.52% was just one quarter of the growth rate in the modern West (Bernstein, 2004: 204). This was followed by more explosive growth of England, several hundred years later related to the first industrial revolution (Bernstein, 2004; Ferguson, 2011). Subsequent industrial revolutions have led to increased economic development. GDP per capita, which can be viewed as an indicator of income, has shown an exponential curve for advanced countries (Danielmeyer and Takeda, 1999) (see Figure 4.1).

Technology has long been considered as influencing economic development (Steenhuis and de Bruijn, 2012). Technology can represent an ability to do new things (effectiveness) and can also represent better ways of doing things (efficiency). AM is an example of a technology that can improve both. New technologies such as AM are the result of innovation. Innovation can be defined as the process of turning ideas into reality and capturing value from them (Tidd and Bessant, 2009). When technology is involved, it means the new development and introduction of knowledge-derived tools, artifacts, and devices by which people extend and interact with their environment (Tornatzky and Fleischer, 1990).

Traditionally, several different types of innovations are distinguished (Marquis, 1969), and the different types of innovations have different impacts

DOI: 10.4324/9781003399100-5

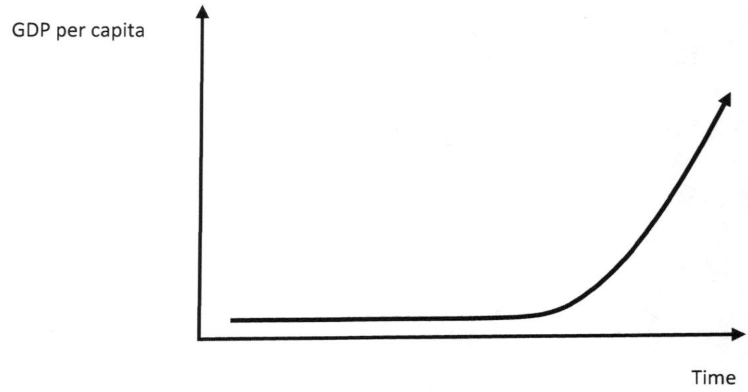

Figure 4.1 GDP per capita for advanced nations.
Source: Based on Danielmeyer and Takeda (1999).

(Steenhuis and Pretorius, 2017). One distinction is that between radical and incremental innovation. A radical innovation provides a brand-new functional capability, which is a discontinuity in then-current technological capabilities. An example of a radical breakthrough in technology that changes the whole character of an industry is the jet engine. Radical innovations are quite rare (Marquis, 1969). Incremental innovation is essential for the typical firm's survival. It relates to ordinary, everyday, within-the-firm kind of technological change that improves the existing functional capability of an existing technology through improving performance. Incremental innovation is paced more by economic factors than radical innovation.

An example of radical AM innovation can be found in the medical industry. AM allows surgeons to print mock-ups that help them to prepare for surgery. This type of preparation is not possible or cost-prohibitive without AM. Donahue (2015) provides a story of a cardiologist who printed a heart mock-up based on the data from CT scans. Once the cardiologist had the mock-up, the medical problem became much clearer and the cardiologist realized that the approach to the procedure would probably have to change from a minimally invasive catheterization to full-blown open-heart surgery. They had prevented a potential complication that was unforeseeable without the AM physical model. An example of an incremental AM innovation is the T25 part that is used inside General Electric (GE) commercial jet engines and that was approved by the Federal Aviation Administration in 2015. The part is a fist-sized piece of silver metal that houses the compressor inlet temperature sensor inside a jet engine. The innovation in this case is the use of an AM process versus a TM process.

Another important distinction is that between sustaining and disruptive innovation (Christensen, 2011). Sustaining technologies tend to maintain a rate of improvement; that is, they give customers something more or better in

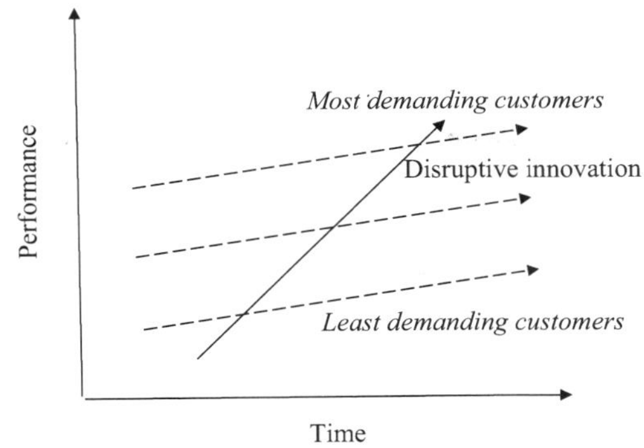

Figure 4.2 Illustration of disruptive innovation.
Source: based on Christensen (2011).

the attributes they already value. This can be in terms of both incremental and radical innovation. The technical definition of a disruptive technology is often misunderstood and does not align with common public perception. A disruptive technology introduces a package of attributes which is very different from the one that mainstream customers historically value, and disruptive technologies often initially perform far worse along one or two dimensions that are particularly important to those customers. This means that disruptive technologies initially underperform and target low-end customers (Bower and Christensen, 1995; Christensen, Raynor and McDonald, 2015). Uber is not a disruptive innovation: Uber did not target low-end customers but instead focused on the mainstream market (i.e., people who were already in the habit of hiring rides). Tesla's approach to electric cars is another example that is not disruptive: it initially targeted the high end of the market with high profit margins (Thomas and Maine, 2019). Disruptive innovation is illustrated in Figure 4.2. Christensen (2011) provides detailed examples of this for the hard disk drive industry and hydraulics technology in the mechanical excavator market. An example of disruptive AM is the application of AM for mass production. The initial application of AM was in rapid prototyping (which is followed to lesser degrees in rapid tooling, direct manufacturing and ultimately household manufacturing; Rayna and Striukova, 2016). AM is generally still considered uncompetitive for mass production; that is, it is performing worse than TM processes. However, as the AM technologies continue to improve, it becomes competitive at higher volumes.

Adoption and diffusion

A technology can show potential but whether that gets realized depends on whether people will use it. Several areas of scientific inquiry are concerned with examining how innovations are adopted by individuals or organizations and how this leads to the diffusion of the technology amongst populations with well-established theories such as the technology acceptance model (Davis, 1989), the technology-organization-environment framework (Tornatzky and Fleischer, 1990), the diffusion of innovation (Rogers, 1995), and the unified theory of acceptance and usage of technology (Venkatesh et al., 2003). Without getting into too many technical details of these theories, the goal here is to provide a general sense of adoption so that how AM technology is spreading can be understood. Rogers (1995) found that people who adopt an innovation can be grouped into five adopter categories based upon when they adopt (see Figure 4.3).

The five adopter categories are explained more in Table 4.1, which explains why some people adopt a technology early whereas others adopt a technology late.

Figures 4.4 and 4.5 show the pattern of cumulative sales of AM machines for, respectively, industrial and desktop 3D printers. This shows that the sales of desktop 3D printers took off much later than that for industrial 3D printers and that sales for both are continuing to increase.

The question is: where is AM in terms of its diffusion? It is clear from these two figures that AM has not reached the late majority. But is AM still in a really early phase or is it already nearing the end of the early majority? To determine this requires more insight into how a high technology gets adopted and diffused. There are different types of adoption, as illustrated in Figure 4.6.

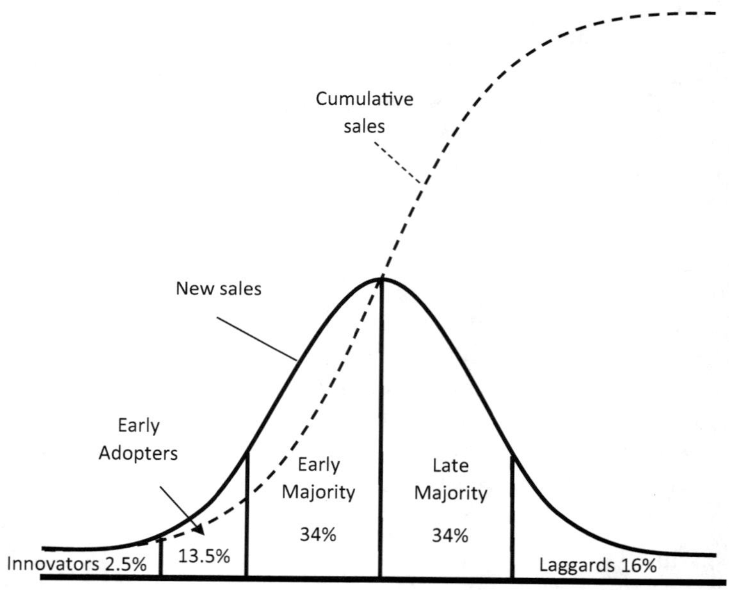

Figure 4.3 Adopter categories.

Source: Adopted from Rogers (1995).

Table 4.1 Characteristics of adopters

Adopter category	*Key characteristics*
Innovators – venturesome	Interested in new ideas. Financial resources that help to absorb possible loss from unprofitable innovation. Ability to understand and apply complex technical knowledge. Connects with other innovators outside of local circle (i.e., more cosmopolite).
Early adopters – respect	Greatest degree of opinion leadership. Potential adopters look to early adopters. Early adopter decreases uncertainty about a new idea by adopting it and then conveying a subjective evaluation to near-peers through interpersonal networks More integrated part of local social system
Early majority – deliberate	Adopts new ideas just before the average member of a system Frequently interact with peers but seldom hold position of opinion leadership May deliberate for some time before completely adopting a new idea. Innovation-decision period longer than that of innovator and early adopter.
Late majority – skeptical	Adopts new ideas just after the average member of a system Adoption may be due to economic necessity and the result of increasing network pressures from peers Innovations are approached with skepticism and caution Their relatively scarce resources mean that most of the uncertainty about a new idea must be removed before they feel it is safe to adopt
Laggards – traditional	Last in the social system to adopt. Almost no opinion leadership Most local-oriented, many are near isolates in their social networks. Traditional values, suspicious of innovations and change agents.

Source: adapted from Rogers (1995).

For illustrative purposes, the focus here will be on the adoption of AM machines by companies to use for production purposes. Note, though, that companies purchase both industrial and desktop AM machines.

Adoption of the high technology: AM

The complication with a technology such as AM is that it can be considered a high technology. While there are difficulties with exactly defining high technology (Hooton, 2018; Steenhuis and de Bruijn, 2006), there are some common elements such as newness, novelty and complexity. These are also characteristics of AM and therefore AM can be considered a high technology.

A key issue with high technology is that it introduces uncertainty. As the technology is still being developed, the manufacturer has to make choices about the technology system but it is unclear how the technology system

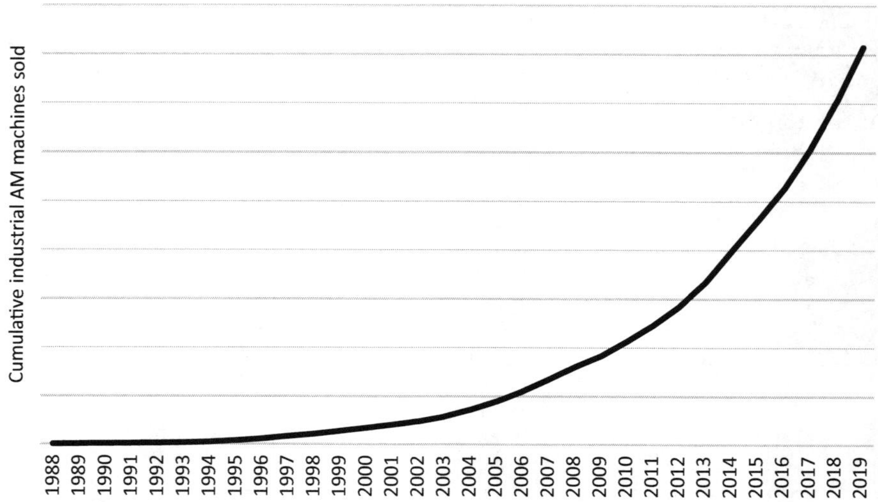

Figure 4.4 Pattern of cumulative sales of industrial AM machines.

Source: Based on sources such as Wohlers and Associated (2020).

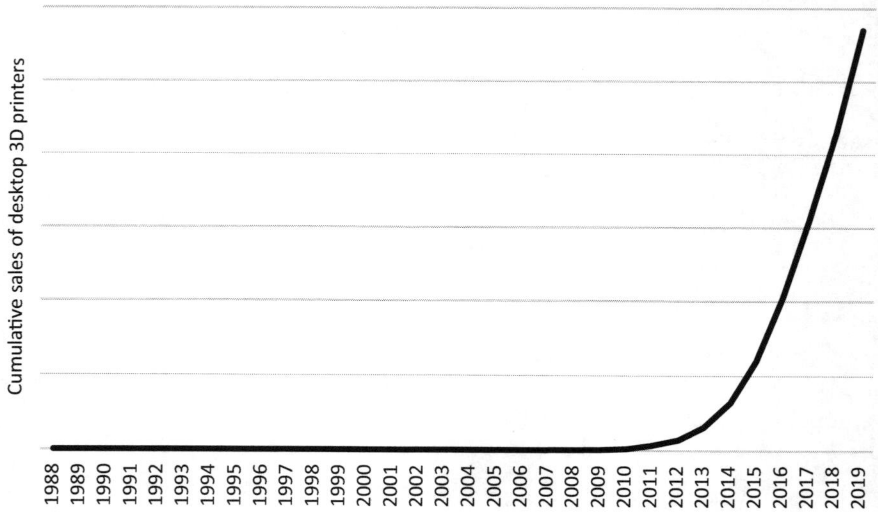

Figure 4.5 Pattern of cumulative sales of desktop AM machines.

Source: Based on sources such as Wohlers and Associated (2020).

characteristics relate to performance (Bender, 1989; Steenhuis, Fang and Ulusemre, 2022). For instance, for MEX, there are decisions about movement, such as what will move—the print head or the build platform—and in what direction? And how will different types of movements influence the

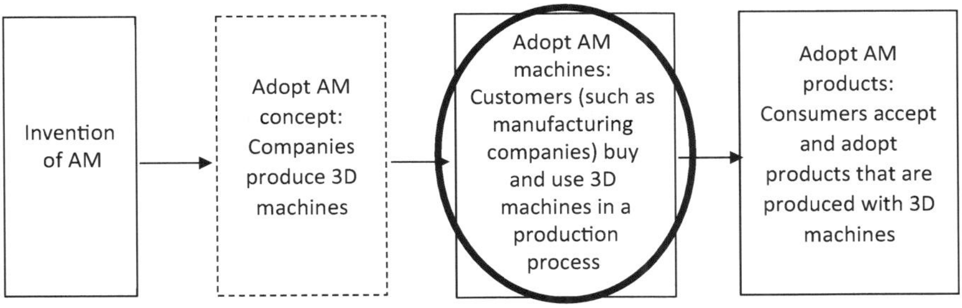

Figure 4.6 Different stages of AM adoption.

Source: Author and adapted from Steenhuis, Fang and Ulusemre (2020).

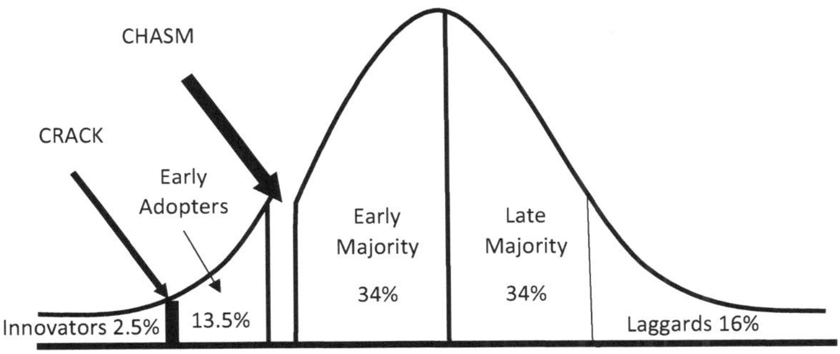

Figure 4.7 Chasm for high-technology adoption.

Source: adapted from Moore (2014).

quality of a 3D printed object? More details on this for desktop 3D printers can be found in Steenhuis, Fang and Ulusemre (2022), but the point is that in high-technology situations where the technology is still undergoing a lot of change, there is high uncertainty for the manufacturer about the relationship between input and output variables as well as what the customer values. For the customer, the unclear relationship between technology system characteristics and performance introduces uncertainty because this lack of clarity makes it difficult to compare 3D printers that have different technology system characteristics. Over time, a dominant design gets established which reduces this uncertainty (Brem et al., 2016). Because of the characteristics of high technology such as uncertainty and change, the model from Rogers as shown in Figure 4.3 has been revised by Moore (2014) to illustrate that there are gaps between adopter categories (see Figure 4.7).

One observed 'crack' is that of a technology that is unable to move from the innovators to the early adopters, such as when a hot technology product cannot be readily translated into a major new benefit (Moore, 2014: 22). The larger gap (i.e., the chasm) occurs between early adopters and early majority

(Moore, 2014: 25). This chasm is defined as the gap between an early market dominated by a few visionary customers to a mainstream market dominated by a large block of customers who are predominantly pragmatists in orientation (Moore, 2014: 6). This chasm can go unnoticed because with both groups the customer list and the size of the order book can look the same. However, what the early adopter is buying is some kind of *change agent*. They expect a radical discontinuity between the old ways and the new, and by adopting the technology, they expect to get a jump on the competition. The early majority, however, is buying a *productivity improvement*. They are looking to minimize the discontinuity with the old ways. Also, by the time they adopt, they want it to work properly and to integrate appropriately with their existing technology base (Moore, 2014: 25). Both adopter categories before the chasm—the innovators (technology enthusiasts) and the early adopters (visionaries)—are technology-oriented and accept risk. But the adopter group after the chasm, the early majority (pragmatists), do not like risk and look for proven concepts that will solve their problems. "When pragmatists buy, they care about the company they are buying from, the quality of the product they are buying, the infrastructure of supporting products and system interfaces, and the reliability of the service they are going to get" (Moore, 2014: 56). Table 4.2 provides some key aspects regarding these adopter categories. One consequence of this is that a signal that the market is dealing with early adopters is when manufacturers are focusing their messaging on the benefits that can give customers an advantage in the marketplace,

Table 4.2 Key aspects of some adopter categories

Adopter category	*Key characteristics*	*Challenge for diffusion*
Innovators	Technology enthusiasts. Technology-oriented and willing to take risk. Can handle certain technology uncertainties.	The technology has to be exciting for the technologists.
Early adopters	Visionaries. Expects radical discontinuity, adopts for change and expects to jump the competition.	Technology must offer the potential of a major change which, when adopted, benefits the adopter and can propel the adopter ahead of the competition.
Early majority	Pragmatists. Adopts for productivity improvement. Looking to minimize the discontinuity with the old ways and wants the technology to work properly and to integrate it appropriately with existing technology.	Technology must be established. There must be proof that it works. There must be knowledge about how it functions, how it will affect productivity, and how it relates to the existing technologies.

Source: Based on Moore (2014).

while when dealing with the early majority, companies will be focused on communicating how established the technology is, the reliability, and so on. Many companies that are selling AM machines focus on communicating the benefits, i.e. the capabilities of AM, in the current marketing and communication. Based on this, it can be concluded that, generally speaking, the industry is at best in the early adopter stage, so that at most 16% of the total potential cumulative sales (and probably significantly less) has been reached.

5 The AM business case II

Across the value chain

This chapter continues the discussion from Chapter 3, where initial insights were provided on the advantages of AM compared with TM. It also continues the discussion of Chapter 4 where information was shared on how innovations are adopted and diffused. In order for AM to spread among companies, there must be a reason for companies to adopt it. This means that AM must be an attractive part of the business model. There is no clear definition of a business model, but based on a review of a variety of sources, Shafer et al. (2005) define a business model as a representation of a firm's underlying core logic and strategic choices for creating and capturing value within a value network. Based on that, this chapter discusses the issue of value, the value chain and where in the value chain AM can add value.

Delivering value

The value chain is a basic tool to understand competitive advantage by examining the activities that a firm performs; it is depicted in Figure 5.1 (Porter, 1985).

The way that a company performs its activities compared with its competitors influences the value that it provides, its value proposition and thus its competitive position. For instance, IKEA has a value chain different from that of other furniture companies. IKEA customers have to assemble furniture themselves, thus reducing the operations cost compared with competitors. In many instances, customers have to pick up the products from an IKEA store, but even in cases of (limited) shipping, the size of the package is smaller and easier to ship because the product is unassembled. Thus, IKEA has lower outbound logistics costs. There are other differences, such as warehousing cost, which can be lower for IKEA as it keeps unassembled products in inventory. The point is that different companies can do things differently to create and capture value. The value chain as shown in Figure 5.1 is for one firm, but firms are linked together, as shown in Figure 5.2. This essentially shows a supply chain. Technically, a supply chain shows how supplies (parts) move through the chain, whereas a value chain shows how value is added in parts of the chain, but there is overlap in these terms.

DOI: 10.4324/9781003399100-6

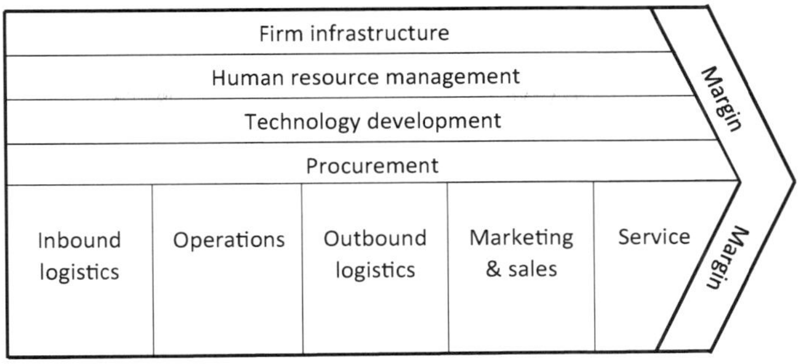

Figure 5.1 The value chain.

Source: adopted from Porter (1985).

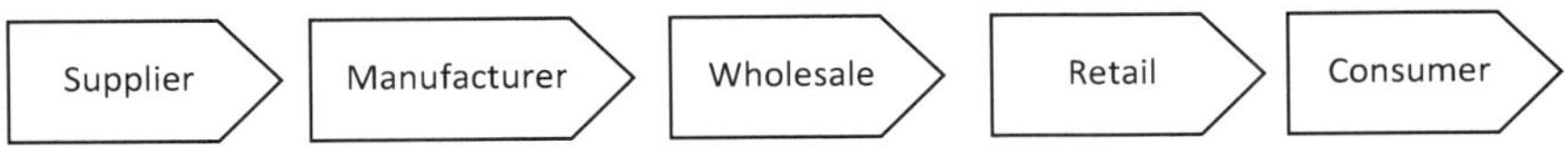

Figure 5.2 General depiction of a supply chain.

Source: Author.

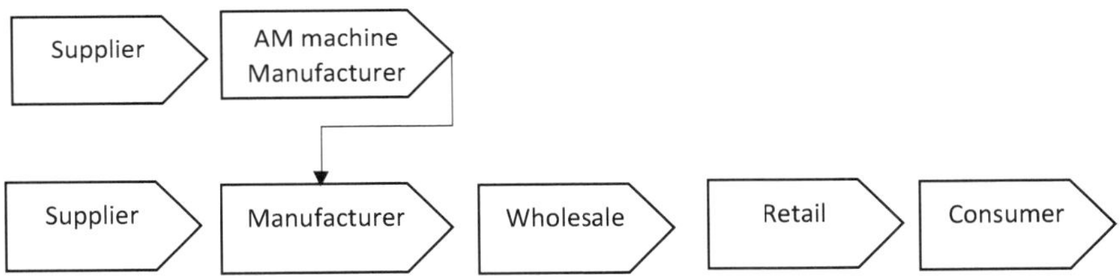

Figure 5.3 Place of AM in a supply chain.

Source: Author.

AM can influence this chain. Figure 5.3 illustrates the position of firms that produce 3D printers in a supply chain. These firms are essentially part of the machine tool industry. That means they produce tools that are used by other firms in their production process. Hence, the output of the company that produces 3D printers (i.e., a 3D printer) becomes an input (i.e., a production machine) for the manufacturer to use in its production process.

When one considers the prosumer perspective, there is a slight change. This is depicted in Figure 5.4. Note that in this case the wholesale and retail parts are no longer part of the chain.

Figure 5.5 presents an expanded version of Figure 5.4. This goes back to the discussion in Chapter 2 which delved into AM and additional technologies that are part of this, such as (a) a design which can be an existing design or something that the prosumer designs themselves (in which case design software

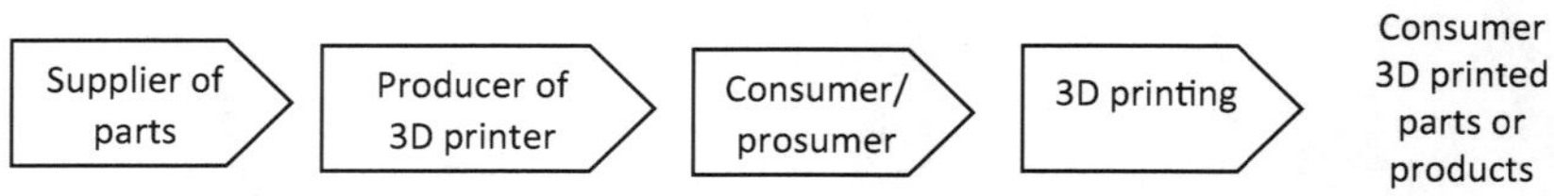

Figure 5.4 Prosumer AM chain.

Source: Author.

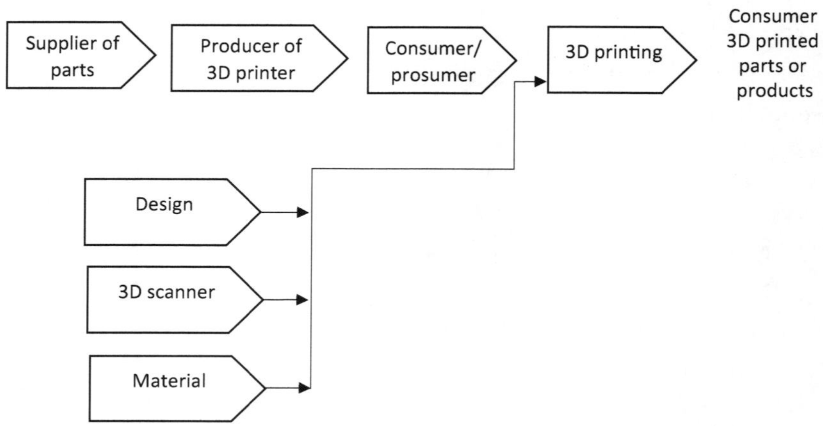

Figure 5.5 Expanded model for prosumer AM chain.

Source: Author.

is needed), (b) the design which is created through a 3D scanning process, and (c) materials for the 3DP process. Prosumers can also sell or post their designs for others to use, but for simplicity's sake, this is not included in Figure 5.5.

There are multiple ramifications of the introduction of AM which follow from Figures 5.1 through 5.5 and changes to the value chain. Some examples are provided below.

1. A traditional manufacturer may adopt an industrial 3D printer to use in its production process because it provides a different value such as lower operations cost (see the discussion in Chapter 3). For instance, Bose used to outsource designing and redesigning parts for their headphones which averaged \$35 per part and required 3 days to complete. Bose then adopted Ultimaker 3D printers, moved the task in-house, and reduced cost to \$1–\$2 per part and just a few hours of 3DP.[1]
2. Owing to the rapid prototyping characteristics of AM, it allows the reduction of uncertainty for entrepreneurs who want to start their own business. Before AM, they may have had to enlist the help of a custom manufacturer to design a part and produce it, which would have been

time-consuming and costly. AM allows for a quick succession of prototypes, which is faster and less expensive.

3. Opening up of retail stores to sell 3D printers. Makerbot, one of the early prosumer-oriented companies, created its own retail stores, although it ended up unsuccessful. Instead of this narrow focus, what seems to work is to offer a broad range of AM-related products and services. An example is the 3D Printer Superstore in Melbourne, Australia. This store offers 3D printers from different manufacturers, scanners, filament, and design software.
4. Similarly aimed at the prosumer, in particular do-it-yourself or makers and perhaps budding entrepreneurs, is the growth of maker spaces and fab labs. Fab labs is a program that started at the Massachusetts Institute of Technology (MIT) and involves small-scale workshops. Maker spaces and fab labs provide a range of tools and machines, including 3D printers. People can go in and use these to make their own products. Especially in education, there is growth in the creation of such spaces. These spaces were previously not possible since TM equipment, such as CNC machines, is expensive and needs a lot of training to operate.
5. A new manufacturing company may come into existence to take advantage of characteristics of AM, such as the ability to produce new products or complex parts. This allows the manufacturing of products that was previously not possible from a competitive perspective. For example, several companies have taken advantage of the rapid prototyping characteristics of 3DP or, to put it differently, the ability to quickly produce unique products. Companies such as 3Dkeepsakes and My 3D Mini Me are examples of companies that utilize scanners and 3D printers to print miniature versions of people and pets. This type of business did not exist before the advent of 3DP as the making of molds was too expensive, especially considering limited quantities.
6. Companies start to flourish in areas such as design; that is, they sell designs by customizing designs for customers or offer a platform for the sharing of designs. This also relates to what companies are offering: (a) products such as 3D printed parts, (b) services such as design, or (c) both. 3D Hubs will custom-design and manufacture parts, TinkerCAD is a (free) online program that allows users to design their own products, and Thingiverse is a free online platform where people can share their designs.
7. Companies start to flourish in areas such as 3D scanners. Shapify has developed booths to scan people, Shining 3D specializes in more regular scanners, and several companies that produce 3D printers have extended their product lines into 3D scanners, such as Makerbot and XYZprinting.
8. Companies start to flourish in areas such as the materials used in AM processes. This extends into companies that develop technology to recycle plastic. Filabot and 3devo are examples of companies that provide machines that can recycle plastic and create filament for 3DP.

9. New companies come into existence because of the further development of the 3DP technology such as specialized parts. Fleks3d is a company that specializes in flexible build plates, Buildtak specializes in 3DP surface material for better adhesion of the first layer, and E3D specializes in hot ends.

All of the examples above not only go beyond the companies that manufacture AM machines but also illustrate how AM is changing the value chains from replacing TM activities to adding new value chain links due to new products or activities. *Any* company in a value chain needs to consider how AM can influence its value chain and what it may mean for their competitive position.

The client order decoupling point and service

Related to the concept of the value chain is the issue of customization. Generally, customization comes at additional cost. The more standard and predictable the production process is, the easier it is to optimize in terms of efficiency. This relates to the issue of process choice (discussed in Chapter 3). AM influences the ability to produce customized products. From a manufacturers perspective, two concepts are important in order to understand customization: (a) the distinction of service from manufacturing and (b) the customer order decoupling point (CODP). These two concepts are related and have to do with what is being done, and where, in the value chain.

Sampson and Froehle (2006) provide a technical distinction between manufacturing operations and service operations. Both are considered extremes, and most companies will operate somewhere in the middle (i.e., will have a mix of service and manufacturing). Service is defined as a process where the customer provides significant inputs into the production process (Sampson and Froehle, 2006). This means that a service process is a process that requires customer input that is necessary for the production of a unit for that customer so that without the customer input, the service process cannot take place. There are three types of customer inputs: customer self (such as during a cab ride), customer belongings (such as shipping a package for a customer), and customer information (such as an accountant preparing a tax return for a customer on the basis of financial information provided by the customer). Manufacturing processes, on the other hand, can take place without a customer input and are thus predictable. For instance, the publishing of a novel happens without customer input. The book is written, then printed and published and then put on display in a bookstore where a customer can buy it. Most of that process is manufacturing and does not require customer input (although usually some type of forecasting is used to get an idea of the customer demand). This makes it relatively easy to manage as it can be determined how many books are going to be printed, when and where and based on that it can be determined how much paper is going to be required and when. Services, on the other hand, require a customer input which introduces an element of uncertainty and reduces efficiency and influences the cost of operations. For instance, West Coast Customs, a company shown in a Netflix series, produces custom-built cars. One customer might want to get a customized VW Beetle, whereas another customer might want to have a customized Ford Mustang. Perhaps these customers already own the basic versions of these cars, or perhaps they need West Coast Customs to provide the cars as well. In any event, it is clear that without the cars, no customization can take place. Furthermore, since West Coast Customs doesn't know in advance whether a customer who wants a customized VW Beetle or a customized Ford Mustang will walk in the door, it is not possible to accurately predict the demand. The same may apply to the custom parts. Since it is expensive to keep inventory of all kinds of cars just for the possibility that a customer might want a customized version of a certain car, the process of customization, which may involve the initial purchase of the basic version of the car, has to wait until the customer arrives and a dialogue about what the customer wants has taken place. From an efficiency perspective, it is better for companies to operate on the manufacturing side of things, whereas from a market orientation, it may be better to operate on the service side of things. AM allows companies to move more towards the service side of things (i.e., customization) without the traditional negative effects on productivity.

The concept of the CODP is related to this as well. The CODP separates the activities that require customization and depend on customer information from the activities that can take place without customization (Rudberg and Wikner, 2004). Figure 5.6 provides an illustration of four different types of processes related to the CODP. An example is Dell, which follows a type of assemble-to-order model. While customers may perceive that Dell customizes laptops, it does so to a very limited degree. Dell keeps many of the parts in inventory so that once a customer places an order, it is merely an issue of

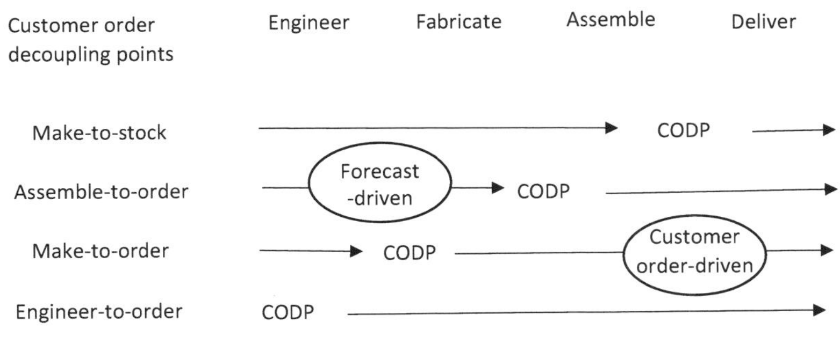

Figure 5.6 Different customer order decoupling points.

Source: Adapted from Olhager (2010).

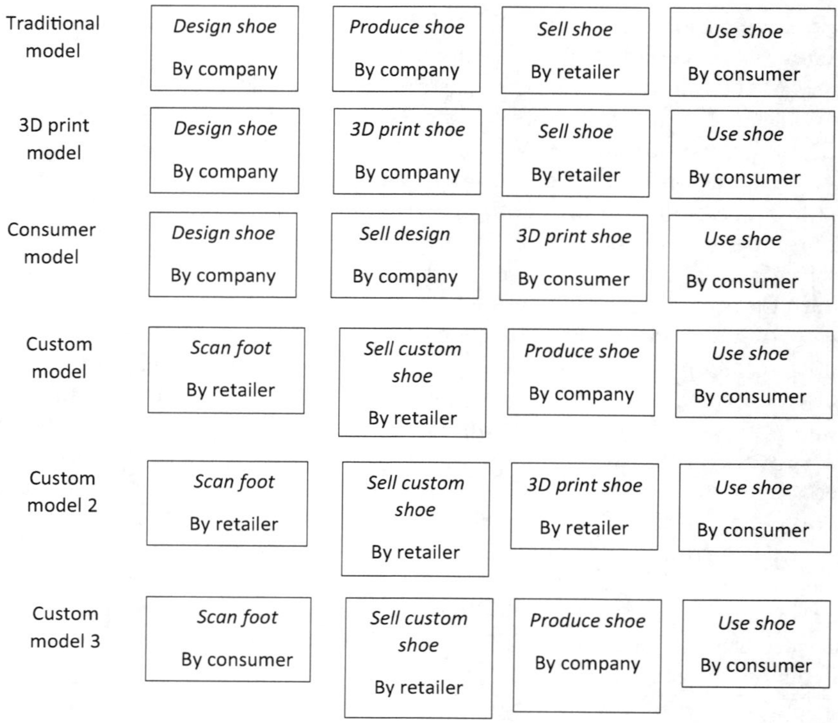

Figure 5.7 Impact of AM on the shoe industry.
Source: Author.

putting the right set of parts together (Parvez et al., 2018). AM changes the situation for manufacturers because of the ability to provide customized products rapidly and efficiently.

Illustrative example

Figure 5.7 provides an illustration of the impact of AM on the traditional value chain related to shoes. It illustrates several alternatives of what activities occur where and by whom to show how the value chain has already changed. With increased technical capabilities in AM, more change is sure to follow.

a) In the traditional model, a shoe company such as Nike designs a shoe with the intent to mass-sell the shoe. The shoe then is typically produced in a low labor-cost country such as India. It is then shipped to retailers in locations across the world. The consumer visits a retail store, can try on the shoe and buys the shoe. Of course, the shopping experience can also be online.

b) A first alternative is that instead of the TM process for shoes, the shoes or parts thereof are produced through 3DP. Examples are athletic shoe companies such as Adidas, Under Armour and Reebok, which have all produced shoes with 3D printed components such as soles. The effect on the value chain is relatively minor.
c) A more drastic alternative to the traditional business model is one where the prosumer starts to take over most of the value chain activities. A company such as 3D shoe provides designs for prosumers. Prosumers buy these designs and then 3D-print the shoes at home. Obviously, this has major consequences for the companies in the traditional value chain!
d) A less drastic alternative is where a customer can get their feet scanned at a retail store so that customized shoes or parts of shoes can be made. These are then manufactured in a central location and from there shipped to the customer. This is done by Superfeet, which offers customized footwear such as insoles. Customers can visit a Superfeet retail location to get their feet scanned. Superfeet uses HP's Fitstation technology for this scanning, which can be locally deployed. This scanning allows for personalized products. Superfeet then produces the customized products at central locations and ships the final product to the customer.
e) Related to the above example is one where the production of the shoes or parts does not take place at a central location but rather is done at the retailer's location with a 3D printer. This requires 3D printers that can work with a variety of materials as well as employees at shoe retailers with technical skills for 3DP. Examples are ECCO and RESA, although it should be mentioned that the 3DP involves only parts: respectively, the midsole and the insole of the shoe (see also Chapter 6).
f) A variation of this theme is where the customer takes care of the scanning part. The company uses this information to provide the customer with a customized design and subsequently produces it. FTSY and vFit offer foot-scanning capabilities. vFit offers a (free) app that allows customers to scan their own feet with their iPhone. This allows a virtual fit. vFit works with a range of retailers, and with the scan information, the customer can virtually try on shoes (i.e., see how their feet will fit inside the shoe) and then decide whether to buy the shoes. With the FitMyFoot smartphone app, customers are able to scan their feet and order tailored insoles.
g) An extreme from the traditional model is the situation where the prosumer takes full control. That means the prosumer designs and then 3D prints the shoes. Showcases of this are Austen Hartley, a Canadian mechanical engineer who designed and customized his own pair of shoes similar to Crocs,[2] and Stephan Henrich, a German designer and architect who designed and printed shoes based on mythical creatures.[3]

These examples show how businesses can have different value propositions and how AM can impact this value. It is not clear which of these business models are the most likely to be successful in the long term.

Notes

1 https://ultimaker.com/learn/bosebuild-accelerating-design-and-testing-phases-with-3d-printed-parts
2 https://3dprintingindustry.com/news/youtuber-combines-speed-and-style-with-custom-3d-printed-crocs-202700/
3 https://3dprintingindustry.com/news/german-designer-unveils-customizable-creature-like-3d-printed-cryptide-sneaker-204957/

Part II

AM in select industries

6 Apparel

The previous chapters have covered the foundation of understanding AM: how it works, why it is appealing for business, how it is spread or adopted, and where in the value chain it is used. Chapters 6 to 10 provide examples of AM applications in several industries, starting with the apparel industry. Defined broadly, apparel here is viewed as including clothes, shoes and also jewelry. By 2020, the worldwide apparel market was estimated at roughly $1.65 trillion.[1] Clothes and shoes can be considered footloose industries. Footloose industries have limited dependency upon proximity to specialized resources and markets. These industries have a lot of freedom in deciding where to locate. For shoe and textile manufacturing, this often means that they are located in low-cost countries. When wages rise, these industries tend to move to other low-cost countries, as shown by Gereffi (1999).

The cost of manufacturing in different locations can be quite different. In 2013, CNN ran a story based on data from the Institute for Global Labour and Human Rights that showed that the cost for making a Denim shirt in the US was estimated at $13.22 whereas in Bangladesh was $3.72. Furthermore, in the US, labor was 57% and material 38% of the total cost; in Bangladesh, the material was 89% and labor only 6% of the total cost.

The cost includes cost related to (environmental) regulations, so that countries with lax laws can be financially attractive. However, several disasters in recent years have called the financial focus into question. In 2013, at a Dhaka garment factory, a building collapsed because of structural failures, killing 1,135 people. Child labor, inequality or poor working conditions can all play a role. In 2017, Nike was faced with protests against so-called sweatshop practices.

Another aspect of the fashion industry is that the sector is recognized as one of the most polluting sectors and a drain on resources (Jacometti, 2019). The fashion industry produces over 92 million tons of waste per year and uses 79 trillion liters of water (Niinimäki et al., 2020). AM has already impacted the apparel industry in terms of how competition takes place and where activities take place (see the example of shoes in Chapter 5). Furthermore, people like Frank Cooper at Birmingham City University have

DOI: 10.4324/9781003399100-8

modernized the School of Jewelry curriculum to include AM for jewelry making. AM also offers the potential to help reduce the negative impact on the climate (see also Chapter 12).

What is being 3D printed in apparel

Some high-fashion items are 3D printed, such as Zac Posen's rose-petal dress (worn by British model Jourdan Dunn at the New York Met Gala), an orchid-inspired dress worn by Miss Singapore during the Miss Universe contest in January 2023, and various designer shoes[2] and designer jewelry.[3] These stunning creations should perhaps be viewed more in terms of haute couture and items for fashion shows rather than as a representation of what is normal for 3D printed items. AM is also used to create special clothes. An example is movie costumes such as a 3D printed crown for the Queen Ramonda character in the *Black Panther* movie. Another example is exploratory work on 3DP of space suits.[4]

For clothes that are more for normal use, designers such as Julia Daviy and Danit Peleg have latched onto 3DP as a means to design and create apparel by experimenting with designs and materials. Danit Peleg, who graduated from Shankar College of Design, created a ready-to-wear collection printed at home by using desktop MEX. She has printed skirts and jackets that can be viewed and bought on her website. These types of clothes can easily be conceived as being worn by people on the street instead of on a fashion runway.

The situation for shoes is a little more complicated. Shoes are assembled from several parts that form the bottom and top of the shoe. The bottom of the shoe consists of outer sole (part that touches the floor), insole (part that touches the feet) and midsole (in between). The top of the shoe is the upper shoe casting (the part that goes over the feet and gets attached to the sole). Companies that use AM have focused mostly on specific parts of the shoe.

- Reebok and New Balance have used 3D printed outsoles; for instance, New Balance has worked with HP.
- Adidas and Under Armour have worked on a 3D printed midsole; for instance, Adidas has worked with Carbon. As a side note, Under Armour sponsored swimmer Michael Phelps during the 2016 Olympics and gave him a pair of 3D printed shoes.
- Dr. Scholl's has 3D printed insoles.
- Nike has printed the upper shoe part. The upper part is normally labor-intensive, and it is more challenging for 3DP because it often consists of multiple materials. Most 3DP processes are limited in terms of printing multiple materials. However, some companies, such as Voxel8, have developed 3DP technology that has multi-material 3DP capabilities. This allows, for instance, 3DP logos and designs on the textile and with varying degrees of stiffness (for support).

- Aside from companies that manufacture parts of shoes, some have produced entire footwear, such as Feetz (although they seem to have stopped operating in 2018).[5] ASICS offers 3D printed sandals, and ATHOS provides customized mountain climbing footwear.
- Tooling, such as shoe molds, has also been produced with AM.[6]

In the case of jewelry, different types of things are being printed. For instance, some jewelry is 3D printed using carbon or plastic.[7] Another use of 3DP is that some process steps in jewelry making can involve casting of wax molds that are subsequently used for the actual jewelry such as rings. An example of a mold for rings that is produced in the traditional way is provided in Figure 6.1. Molds are created for each ring which then are combined into the tree, as shown in Figure 6.1. This type of wax tree can be produced much faster with AM. Yet another use of 3DP is that of printing metal jewelry, in particular of complex shapes that are otherwise difficult or impossible to make. Some great examples with pictures can be found at https://all3dp.com/3d-printed-jewelry/.

Figure 6.1 Mold for making rings.

Source: author.

Performance: why AM printing is used for apparel

The reasons why AM is used for apparel go back to the discussion in Chapter 3 and, in particular, on the capabilities that are offered by AM and how they compare with TM (see Table 3.3).

A main reason for using AM to create apparel is that it allows customization and personalization. This relates to *product flexibility* and *product line flexibility*. Customization plays a role for the high-fashion end of things, but it also relates to regular consumer items as it means that the perfect fit for clothes and shoes can be created on the basis of who is going to wear it or, for jewelry, to get something made to the desire of the customer. Customization may involve luxury items for which a premium price is charged.[8]

Speed is another reason to use 3DP, meaning the ability to deliver in a short time frame (*delivery time*) and a short time frame from product development to market introduction (*time to market*). This applies particularly to 3DP shoes (Sole et al., 2017). Using AM eliminates some steps in the shoe production process, potentially leads to production on location, and overall shortens the supply chain. For customized products for a particular customer, this can quickly lead to the final product. In the case of products meant for larger production volumes, AM can shorten the overall process by a faster design through rapid prototyping. For instance, Heron Preston, a fashion designer, partnered with Zellerfeld and created HERON01 sneakers. AM allowed him to design and print, in hours, functional and evolving protypes which with TM would have taken months.[9] Aside from that, Zellerfeld creates a unique pair of footwear that is specifically for that customer, such as for Ye, aka Kayne West, at the 2022 Paris Fashion week.[10] Similarly, Crocs, a company that offers comfortable leisure footwear, has partnered with Italian 3DP company 3ntr, which allowed it to improve the prototype phase.[11]

Cost also plays a role, namely to provide the product at lower cost compared with TM (*cost efficiency*). In particular for shoes, one of the components of cost is the mold for making shoes.[12] For traditional shoe making, basically, for each type of shoe and size of shoe, a mold is required. AM does not require such a mold. It can therefore reduce cost (and be faster) to produce shoes with 3DP. This also applies to jewelry that is made by using wax molds which can be replaced with 3D printed molds, which is faster and reduces labor.

Innovation, particularly the introduction of new products (*product innovation*), is also prevalent for apparel as a number of designers such as Danit Peleg and companies such as Feetz have created new products. Furthermore, for shoes, a number of different products of a variety of soles have been created. New services, such as scanning feet for optimal fitting of shoes (*service innovation*), have been added as well.

Lastly, AM helps with *process sustainability* and *product sustainability*. AM creates less waste, which is also important in the making of clothes and shoes. For example, Rik Olthuis, a New Zealand design engineer, created a pair of completely biodegradable athletic shoes,[13] and Lucie Trejtnarová

Table 6.1 Why AM is beneficial in the apparel industry

Strategic dimension	*Manufacturing capability*	*Apparel*
Cost	Cost efficiency	For shoes and jewelry, reduction of cost by using AM for tooling
	Resource efficiency	Not applicable
Quality	Product quality	Products do not necessarily have better performance other than having a better fit (part of product flexibility)
Time	Delivery time	Particularly for shoes and for apparel in general depending upon what is produced and where in the value chain
	Time to market	Particularly for shoes because of rapid prototyping
Flexibility	Product flexibility	Customization of clothes, shoes and jewelry
Innovation	Product innovation	New designer clothes and new types of shoes
	Service innovation	New services such as scanning of feet for optimal fit
Sustainability	Product sustainability	Use of biodegradable material such as for shoes
	Process sustainability	Reduction of waste and CO_2 emissions (e.g., for making shoes).

Source: author.

developed biodegradable 3D printed sandals.[14] Modern Synthesis, a company from the UK, is in the early stages of working on a biomaterial for a microbial weaving process which could help the fashion sector to lower its carbon footprint and reduce waste.[15] Furthermore, a team of graduate students at Yale's Center for Business and the Environment, worked with a company called HILOS and several of its supply chain partners (BASF, HP, and AMT) to examine its environmental impact and found that shoe production with AM reduces CO_2 emissions by 48% compared with traditional methods of shoe making.[16] Similar to Table 3.3, Table 6.1 provides an overview of the strategic dimension and manufacturing capabilities related to AM that affects the apparel industry in particular.

How apparel is 3D printed: use of AM technology

The main AM technologies used in apparel are MEX, VPP and PBF (Kim et al., 2019). For example, MEX was used to 3D print high-heel shoes for luxury brand Balenciaga,[17] VPP was used to 3D print Adidas midsoles, and PBF technologies are used to make metal jewelry (Cooper, 2016). MJT and BJT are less prevalent but have been used for 3DP of clothes (Vanderploeg, Lee

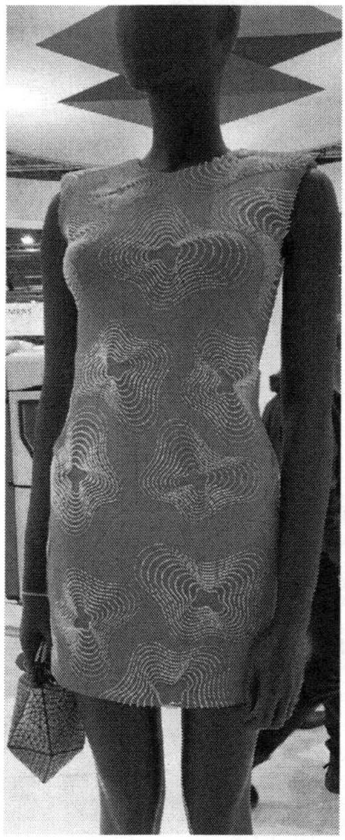

Figure 6.2 3D printed designs on clothes.
Source: author.

Table 6.2 AM technologies used for apparel

	AM process						
	MEX	*PBF*	*VPP*	*MJT*	*BJT*	*DED*	*SHL*
Apparel	Most commonly used			Less prevalent		Uncommon	

Source: author.

and Mamp, 2017). 3DP of designs on clothes is illustrated in Figure 6.2, which was created by Stratasys. However, MJT and BJT are less prevalent. Table 6.2 provides a summary.

Where apparel is 3D printed: AM in the apparel value chain

AM is applied throughout the value chain from the research and development (R&D) and supply side to the consumer side and many of the steps in between. See the discussion in Chapter 5 that provides the various approaches related to shoes.

Diffusion and challenges of AM in apparel

In apparel, in terms of Figure 4.6, the concept of AM has been accepted, some specialized AM machines have been made and companies have adopted the machines for production purposes. To some degree (e.g., with shoes), the AM concept has been adopted in the value chain, and products produced with AM have been accepted by consumers as well. However, for the overall adoption in rough terms (specific products can be in slightly different stages), it seems that AM is not widespread and is, in terms of Figure 4.7, dealing with the early market (i.e., innovators and perhaps the early adopters for customized apparel and shoes).

Using AM for the manufacturing of clothes and shoes faces at least two main challenges: comfort and the use of (a mix of) materials (Hohn and Durach, 2021). Shoes and clothes need to fit comfortably, and the material needs to breathe, be durable, and move freely (e.g., for clothes when sitting down) (Perry, 2018). Without these characteristics, customers are unlikely to buy 3D printed clothes and shoes. Existing AM technologies are already capable of printing products that are quite flexible (see Figure 2.14), and flexible materials such as FilaFlex, used by Danit Peleg, also exist. Nevertheless, it is not yet the same as regular clothes, but there is continual effort to improve materials in terms of softness and flexibility.

For shoes and clothes, additional challenges are the weight of the material, the potential need for polishing and thus post-processing, and whether the clothes are washable (Perry, 2018). All of these relate to materials, so the future success of 3D printed clothes and shoes depends largely on what type of advancements will be made. AM may also provide the potential to use materials that cannot be used in TM to produce clothes. An example is a material that can be used in 3DP, allows the body to keep cool in warm temperatures, and is significantly better than some alternatives such as cotton.[18]

Cost is a major challenge for 3DP of jewelry. For instance, metal jewelry can be 3D printed, but unless it has complex shapes that are difficult to make with TM, it is unlikely to be cost-competitive (Cooper, 2016).

An additional, but somewhat different, challenge is related to the business model. Especially in a new and developing area, there are people with great ideas, but this doesn't always lead to a financially sustainable business. Continuum Fashion, which 3D printed the first ready-to-wear bikini, seems to be defunct, as is Feetz, the company that 3D printed shoes. Even Adidas, with its hyped up Speedfactory, which was meant to bring shoe production back to Europe, ended up closing the factories and moving production back to Asia. The closure indicates that, while successful in areas such as the ability to customize and automate, it is still too costly to operate close to a market such as Europe. In other words, despite reducing delivery lead times to stores, this didn't provide enough additional value to offset the additional infrastructure cost and workforce cost. There are still too many steps involved in making shoes and these cannot all be automated. Thus, there is still (too much) dependence on (high-cost) labor to make this viable in high-labor-cost countries.

Notes

1 https://www.statista.com/statistics/551775/size-of-the-global-apparel-and-footwear-market/
2 http://www.designgallerist.com/blog/3d-printed-shoes-by-zaha-hadid-and-more-famous-designers/
3 https://3dprint.com/33147/3d-printed-jewelry-dijkstra/
4 https://3dprint.com/289574/nasas-futuristic-projects-3d-printed-spacesuits-and-micro-robot-swarms/
5 https://www.futurefootwear.in/3d-printed-wearable-shoes-feetz-com/ shows how shoes are printed
6 https://www.fabbaloo.com/news/a-metal-3d-printer-for-shoe-printing?
7 https://inhabitat.com/3d-printed-jewelry-company-uses-plants-not-fossil-fuels-to-make-its-beautiful-designs/
8 https://3dprint.com/293872/luxury-giant-richemont-pursues-3d-printed-watches-via-panerai-brand/
9 https://3dprint.com/285957/fashion-designer-reimagines-footwear-landscape-with-3d-printed-heron01-sneaker/
10 Apparently, customers are required to apply to be beta testers for the company's shoes before they're even given permission to purchase them https://3dprint.com/294767/ye-debuts-3d-printed-boot-powered-by-zellerfeld-for-paris-fashion-week/
11 https://www.fabbaloo.com/2020/10/crocs-and-3d-printing
12 Examples of what these molds look like can be found on the internet. See, for example, https://www.pinterest.com/pin/16325617386081855/
13 https://3dprint.com/273641/new-zealand-designer-wins-national-james-dyson-award-for-sustainable-3d-printed-shoes/
14 https://3dprintingindustry.com/news/lucie-trejtnarova-and-fillamentum-create-biodegradable-3d-printed-sandals-159957/
15 https://3dprintingindustry.com/news/modern-synthesis-raises-4-1-million-to-increase-fashion-sustainability-with-textiles-grown-from-microbes-210695/
16 https://hilos.co/pages/hilos-case-study-hilos-case-study-the-first-environmental-assessment-of-3d-printed-footwear
17 https://www.fabbaloo.com/news/the-tech-behind-those-3d-printed-balenciaga-shoes
18 https://phys.org/news/2017-11-fabric-cool.html

7 Construction

The construction sector is one of the largest in the world economy; about $10 trillion is spent on construction-related goods and services each year (Barbosa et al., 2017). The industry's productivity has not improved as much as that of many other industries in recent decades, and AM offers an opportunity to change this (Perrot, 2019). The construction industry has also long been a leader in waste production. In terms of resource consumption, it is estimated that 40% of all raw materials used globally are used in the construction industry, and an estimated three to four tons of waste are generated during the construction of a typical single-family home (Perkins and Skitmore, 2015: 5). Producing a ton of cement requires 4.7 million British thermal units (BTUs) of energy, which is equivalent to about 400 pounds of coal, and this process generates almost a ton of CO_2.[1]

What is being 3D printed in construction

AM is used for the construction of houses, and the media frequently reports on the first 3D printed house. For instance: 'First 3D printed house built in Amsterdam' (in May 2014),[2] 'WinSun China builds world's first 3D printed villa and tallest 3D printed apartment building' (in January 2015),[3] 'World's first 3D Printed building in Dubai' (in May 2016),[4] 'Apis Cor 3D prints first on-site house in Russia in one day for $10,134' (in March 2017),[5] 'First 3D printed house built in Yaroslavl' (in October 2017),[6] 'The first permitted 3D-printed home unveiled' (in April 2018),[7] 'World's first 3D printed house constructed in Belgium' (in July 2020)[8] and 'Construction begins on CyBe's first 3D printed homes in the Caribbean' (in October 2021).[9] The difference in these announcements relates to the underlying technological aspects such as material used, size of the 3D printed building, and whether 3DP takes place on site. While these headlines give the impression that entire houses are 3D printed that are ready for people to move into, this is not the case. AM is used primarily for the construction of walls, floors and ceilings, but the installation of plumbing, electric wiring and so on is not part of the process.

In addition to being used in the 3DP of single homes, AM has been used to print an apartment building,[10] concrete pedestrian[11] and cycling bridges,[12]

DOI: 10.4324/9781003399100-9

Figure 7.1 Example of concrete 3D printed stairs by BetAbram in Krko, Slovenia. Featured in the picture are Joze Abram (seated), CEO of BetAbram, and the author (standing).

Source: author.

steel bridges,[13] a small castle,[14] post office,[15] schools,[16] mosque,[17] sets of houses built from 3D printed plastic,[18] and even entire neighborhoods in Mexico,[19] Kenya[20] and the US.[21] While large structures get a lot of attention, AM construction technology is also used for the production of smaller things like a bus stop,[22] planter boxes and moldings,[23] stairs (see Figure 7.1), swimming pools,[24] for the exploration of new types and unique forms of architecture,[25] new types of materials in construction such as salt,[26] hempstone[27] and cellulose,[28] and design such as of highways.[29] AM in construction is applied in developed and developing nations such as Guatemala[30] and Malawi[31] and is planned for use in space.[32]

Performance: why AM printing is used for construction

The key issue for the construction industry is that it has been left behind in terms of productivity improvements. AM provides an opportunity to bring the construction industry into the 21st century through automation. This automation has several effects in terms of manufacturing capabilities.

A key part is sustainability, particularly *process sustainability*. Eighty percent of the total worldwide waste is generated in the construction industry, and there are significant contributions from formwork. Formwork is typically made of wood and is the temporary structure in which wet concrete is poured. It represents 35%–60% of the overall construction cost and is a significant source of waste because it gets discarded. An advantage of the use of AM in construction is the elimination of formwork and thus the reduction of waste and associated cost (i.e., *cost efficiency*) (Holt et al., 2019; Sanjayan and Nematollahi, 2019). *Product sustainability* takes a role in, for example, use of 3D printed foam for isolation material, which can help reduce CO_2 emissions.[33]

Reduction of waste also leads to lower cost (*cost efficiency*), while another advantage of AM in construction is labor savings. 3D printers allow a house to be built by a skeleton crew rather than a full team spanning multiple trades (Holt et al., 2019). Whether the lower cost translates to a lower price for consumers remains a question mark. A large variety in prices for 3D printed houses are reported.[34]

Another advantage is reduction of time (*delivery time* and *time to market*). In Chapter 3, it was explained that AM processes are typically slow compared with TM but the exception to this is construction. A study has shown a 60% reduction in construction duration compared with the use of conventional methods of construction (Malaeb, AlSakka and Hamzeh, 2019). A practical example is WinSun, which claims that their AM construction requires 60% less materials than typically needed to construct a home, they can be printed in a time span of 30% of that of traditional construction and they need 80% less labor.[35] Also playing a role is quality (*product quality*), such as improved ability to withstand natural disasters,[36] fire resistance[37] and the claim that a 3D printed building can withstand an earthquake measuring 8 on the Richter scale.[38] The quality is dependent upon the material used, so several companies that developed construction 3D printers also developed their own proprietary materials to be used in those 3D printers.

Safety is another reason to use AM in construction. Safety is a bit tricky in terms of the manufacturing capabilities discussed in Chapter 3 but, in the context of construction, fits best under the cost dimension as *resource efficiency* (i.e., the ability to maximize the utilization of process resources [machinery and human]). Western countries in general have higher levels of safety regulations than developing countries, but studies in Australia and the US have shown that, despite this, the injury rates in the construction industry are roughly 4%–5%. This makes it among the highest injury rates when compared with other sectors. A benefit of automation is the potential to meaningfully decrease the number of injuries and deaths in the construction sector by avoiding many of the dangerous tasks (Sanjayan and Nematollahi, 2019: 9).

Additional advantages of 3DP in construction include reduction of air pollutants emission (*process sustainability*), use of eco-friendly materials (*product sustainability*), reconstructing classical architecture (*product flexibility*),

Table 7.1 Why AM is beneficial in the construction industry

Strategic dimension	*Manufacturing capability*	*Construction*
Cost	Cost efficiency Resource efficiency	Lower cost due to less waste and less labor Better efficiency due to better safety
Quality	Product quality	Better construction such as ability to withstand disasters
Time	Delivery time Time to market	AM is faster than traditional construction
Flexibility	Product flexibility	Customization of buildings with novel load-bearing geometries and new types of materials
Innovation	Product innovation Service innovation	This does not seem to be a distinguishing factor compared with TM.
Sustainability	Product sustainability	Use of 3D printed foam can help insulate houses better, leading to lower CO_2 emissions.
	Process sustainability	Reduction of waste in the construction process

Source: author.

and increase in architectural freedom (*product flexibility*) (see Oberti and Plantamura, 2015; Perkins and Skitmore, 2015; Rouhana et al., 2014; Urbanic et al., 2013). Innovation does not seem to be a distinguishing factor as both traditional and AM construction are involved in low unit volumes with a high degree of customization. However, with AM, it is possible to create unique shapes (which falls under product flexibility). Similar to Table 3.3, Table 7.1 summarizes the above discussion in terms of how AM affects the strategic dimension and related manufacturing capabilities for the construction industry in particular.

How: AM technologies used in construction

Two AM technologies are most widely used in construction: BJT and MEX (Marchment et al., 2019; Nematollahi et al., 2019; Nerella and Mechtherine, 2019). BJT is suitable for small-scale building components that can be assembled on site and is a method capable of producing building components with fine details and intricate shapes (Xia, Nematollahi and Sanjayan, 2019a) that can be competitive with concrete (Elsayed et al., 2022). An example of the use of BJT in construction comes from D-Shape (Avrutis, Nazari and Sanjayan, 2019; Feng et al., 2019). D-Shape is a company with early involvement in 3DP. Its 3D printer uses first a layer of sand and, immediately after, a layer of binder (glue), turning the sand into a solid concrete-like object.

MEX is also commonly used where some form of concrete is poured into a pattern. Two different approaches are followed: on-site and off-site

Table 7.2 AM technologies used for construction

	AM process						
	MEX	*PBF*	*VPP*	*MJT*	*BJT*	*DED*	*SHL*
Construction	Most commonly used	Uncommon			• Most commonly used	Uncommon	

Source: author.

construction. An example of the latter is WinSun. It uses off-site machines that are 6.6 meters high, 10 meters wide and 150 meters long.[39] These are used to produce walls which are then shipped and assembled on site. An example of on-site construction is Contour Crafting. Its equipment can be moved to a construction site to 3D print on location. Table 7.2 provides a summary.

Where AM is used in the construction value chain

Research and development are steps before the actual construction. A variety of universities across the world are involved in research on the technical aspects of 3DP in construction. Examples are the University of Eindhoven from The Netherlands[40] and Loughborough University from the UK.[41] Particular attention is given to the topics of materials[42] and sustainability.[43] Construction in space is another major area. In 2022, NASA awarded ICON a $57.2 million contract for lunar structures.[44] The same year, a 3D printed space habitat was opened for students to live in for an immersive learning experience. This was made by the Swiss Institut auf dem Rosenberg in collaboration with SAGA Space Architects and constructed to fit into SpaceX's Starship rocket.[45] The motivation for construction in space can easily be understood: the cost of transporting materials into space is prohibitive – potentially $2 million for a single brick to be shipped to the Moon (Leach, 2014). AM may provide a solution to the problem of construction in space in two ways. First, it may allow for shipping raw materials (taking up less space). Second, it may allow for materials found in space to be used in AM processes, reducing the need to ship materials. A study by Warren et al. (2022) used BJT with (simulated) moon and mars materials to demonstrate feasibility for making building materials. With continuing space exploration, AM is therefore considered for missions such as going to Mars (Kading and Straub, 2014).[46]

Some companies in the construction value chain are primarily design-oriented (architects) and have started to incorporate AM technology because of its design advantages (e.g., DUS from The Netherlands). An extension of this is companies that utilize AM to offer customized construction, such as Azure, which specializes in accessory dwelling units (ADUs) and backyard studios. ADUs are smaller, independent houses located on the same lot as the primary

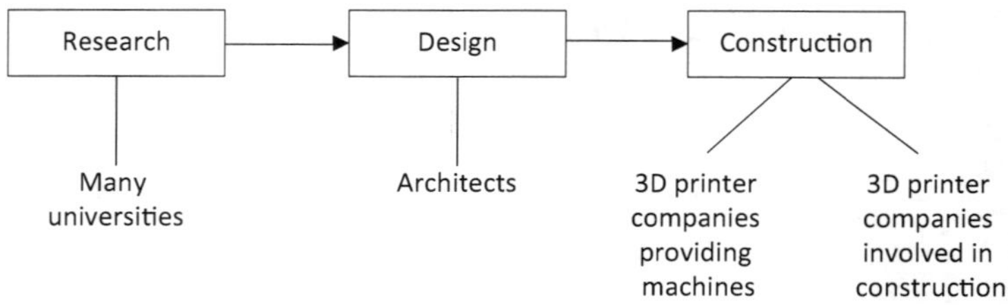

Figure 7.2 AM in construction.

Source: author.

home and can, for example, be rented out to create income. Azure works with the customer to create a design per customer specifications and then 3D prints the unit.

There are also companies that have more emphasis on the development and sales of AM equipment for the construction industry. Examples are US-based Contour Crafting, The Netherlands-based CyBe, Cobod from Denmark, WASP from Italy, Peri 3D Construction from Germany, XtreeE from France, and BetAbraham from Slovenia. A slightly different business model is that of Black Buffalo (in the US), which rents out its machines.

Since the field is relatively new, some companies that developed AM technology are also actively involved in the building and construction as this demonstrates the productivity and feasibility of the technology. Examples of these types of companies are Winsun in China and Apis Cor, ICON and Mighty Buildings from the US.

Lastly, there are situations where the emphasis is on the use of AM technology through a variety of means. For instance, the United Arab Emirates, particularly Dubai, has explicit plans for AM in construction and wants to 3D print 25% of buildings by 2030.[47] They use a variety of companies such as WinSun and CyBe to accomplish this goal. Figure 7.2 summarizes the above discussion and shows where work on AM is taking place in construction.

Diffusion and challenges of AM in construction

In terms of Figure 4.6, the concept of AM has been accepted for construction, specialized AM machines have been made and companies have adopted the machines for production purposes. The AM concept has been adopted in the value chain, and products produced with AM have been accepted as well, meaning that people already live in AM-constructed houses. However, in regard to the overall adoption in rough terms, it seems that AM is not all that widespread and, in terms of Figure 4.7, has not crossed the chasm to the early majority.

While the use of AM technology in construction keeps expanding globally, there are several challenges that influence the rate of adoption. As shown in Chapter 4, it can be challenging to adopt a new and unproven technology.

The construction industry is a traditional and well-established industry. In this type of environment, it is challenging to introduce new technologies, which do not have a long and proven track record. Owing to safety and other concerns, the construction industry is heavily regulated, and it takes time to get regulation passed for a new technology. In 2022, Montana became the first US state to approve 3D printed walls as a fully fledged replacement for those traditionally made.[48]

Key technological constraints for AM in construction are the strength, safety and characteristics of the building materials. Concrete as a material is susceptible to the effects of weather, ambient temperature, and deficiencies in mix design (Holt et al., 2019). 3DP concrete does not yet have established and relevant guidelines or proposed procedures (e.g., for the concrete mixture evaluation) or any set of well-defined acceptance criteria (Kazemian et al., 2019). Furthermore, the use of AM in construction is faced with several conflicting material requirements. With MEX, rheology is an issue: the material has to be soft enough so that it can be pumped to the extruder (pumpability) and poured (printability), but once it is poured as a layer, it has to be hard enough so that it retains its shape (Nerella and Mechtherine, 2019; Perrot, 2019). Moreover, as layers are added upon layers, the layers below need to be strong enough to carry the weight of the layers above (buildability). Buckling (i.e., curving or collapsing of a wall) is a major concern (Suiker, 2018). Reinforcement is used in conventional construction in the form of metal structures around which concrete is poured, which is particularly important for multistory buildings. This is challenging when AM is used, and research is taking place on alternatives for reinforcement, such as on adding fibers to improve the material properties (Hambach, Rutzen and Volkmer, 2019; Nam et al., 2019; Perrot, 2019) and AM methods that 3D print concrete onto a metal rebars net (Perrot, 2019). Interestingly, for some of the mechanical characteristics and some areas of material strength, AM shows better results than conventional pouring of concrete. For instance, the latter is more likely to include trapped air (Sonebi, Amziane and Perrot, 2019). There is also the issue of the printability window, which is the period of time during which the printing mixture can be extruded by the nozzle with acceptable quality, considering the workability loss that occurs over time (Kazemian et al., 2019). The weak interlayer strength between printed layers using MEX is another limitation of the use of AM when compared with cast-in-the-mold concrete (Nematollahi et al., 2019; Xia, Nematollahi and Sanjayan, 2019b).

The issues discussed above cause uncertainty for construction companies that are considering purchasing 3D printers. The business model is unclear (and the payback period or return on investment), and regulatory bodies can become obstacles (although in locations such as Dubai, regulatory bodies enhance the adoption of AM). The uncertainty explains why some companies that have developed construction AM technology are involved in construction itself to demonstrate the technology. This illustrates that the early majority has not been reached. It also explains a focus on smaller architectural components because this can help to establish and diffuse the technology. The companies that have adopted AM construction see it as a disruptive technology that will give them an edge in the market, which is characteristic of early adopters.

Notes

1 https://3dprint.com/80704/3d-printed-hempstone-house/
2 https://www.floornature.com/first-3d-printed-house-built-in-amsterdam-9509/
3 https://www.3ders.org/articles/20150118-winsun-builds-world-first-3d-printed-villa-and-tallest-3d-printed-building-in-china.html
4 https://gulfnews.com/uae/government/worlds-first-3d-printed-building-in-dubai-1.1833450
5 https://www.3dprintingmedia.network/apis-cor-3d-prints-first-site-house-russia-one-day-10134/
6 https://3d-expo.ru/en/article/v-yaroslavle-postroili-perviy-giloy-3d-pechatniy-dom-77606
7 https://www.sothebysrealty.com/extraordinary-living-blog/the-first-permitted-3d-printed-home-unveiled
8 https://www.brusselstimes.com/news/business/technology/120303/worlds-first-3d-printed-house-constructed-in-belgium/
9 https://3dprintingindustry.com/news/construction-begins-on-cybes-first-3d-printed-homes-in-the-caribbean-197543/
10 https://www.cnet.com/news/worlds-first-3d-printed-apartment-building-constructed-in-china/
11 https://3dprint.com/270883/guinness-certifies-worlds-longest-3d-printed-concrete-bridget-in-china/
12 https://www.smithsonianmag.com/smart-news/3-d-printed-cycling-bridge-unveiled-netherlands-180965330/
13 https://www.dezeen.com/2018/10/22/worlds-first-3d-printed-steel-bridge-completed-mx3d-technology/
14 https://www.startribune.com/shorewood-engineer-builds-castle-with-3-d-concrete-printer/273410261/
15 https://www.indiatimes.com/news/india/india-first-3d-printed-post-office-bengaluru-578083.html
16 https://3dprintingindustry.com/news/indonesia-targeting-concrete-3d-printed-schools-after-conducting-initial-trials-of-the-technology-200975/
17 https://www.3dnatives.com/en/3d-printed-mosque-dubai-16012023/#!
18 https://3dprintingindustry.com/news/reinhabit-selects-azure-to-3d-print-ten-new-homes-from-recycled-plastics-in-california-215317/
19 https://www.wbur.org/hereandnow/2020/02/06/worlds-first-3d-printed-neighborhood-mexico
20 https://3dprint.com/297977/cobod-machines-3d-printing-a-house-a-week-in-kenya/
21 https://www.goodnewsnetwork.org/first-3d-printed-neighborhood-is-providing-100-sturdy-affordable-solar-homes-near-austin-for-400k/
22 https://www.3dnatives.com/en/3d-printed-bus-shelters-020120184/
23 https://3dprint.com/293291/legacy-aussie-building-firm-transitions-to-3d-printed-construction/
24 https://3dprint.com/292427/worlds-first-3d-printed-fiberglass-swimming-pool-debuts-on-fox-and-friends/
25 https://3dprint.com/137979/infinite-landscape-house/

26 https://3dprint.com/6974/3d-printed-house/
27 https://3dprint.com/80704/3d-printed-hempstone-house/
28 https://3dprint.com/76838/3d-printed-cellulose-houses/
29 http://scitechconnect.elsevier.com/3d-printing-revolutionizing-highway-design
30 https://www.3dnatives.com/en/3d-printing-housing-shortage-guatemala-200920224/
31 https://www.weforum.org/agenda/2021/06/3d-printed-home-african-urbanization/
32 https://3dprintingindustry.com/news/icon-awarded-new-57-2m-nasa-contract-to-develop-system-than-can-3d-print-lunar-structures-218125/
33 https://3dprintingindustry.com/news/tu-graz-looks-to-revamp-construction-sector-with-concrete-3d-printing-technology-210000/
34 https://all3dp.com/2/3d-printed-house-cost/
35 https://3dprint.com/38144/3d-printed-apartment-building/
36 https://3dprintingindustry.com/news/indian-army-unveils-its-first-two-storey-disaster-resilient-dwelling-unit-220076/
37 https://www.3dnatives.com/en/emergent3d-3d-printed-fire-resistant-houses-california-091220224/#!
38 Durable enough to withstand an earthquake measuring 8.0 on the Richter scale
39 http://www.winsun3d.com/En/News/news_inner/id/402
40 https://www.tue.nl/en/research/research-groups/structural-engineering-and-design/3d-concrete-printing/
41 https://www.lboro.ac.uk/enterprise/3dcp/
42 https://3dprint.com/296325/us-army-corps-of-engineers-awards-3-2m-for-3d-printing-ultra-high-performance-concrete/
43 https://3dprintingindustry.com/news/researchers-show-that-3d-printing-homes-make-them-more-energy-efficient-213410/ and https://3dprint.com/293461/are-3d-printed-houses-as-sustainable-as-they-seem/
44 https://www.nasa.gov/press-release/nasa-icon-advance-lunar-construction-technology-for-moon-missions
45 https://3dprint.com/293151/3d-printed-space-habitat-now-open-to-swiss-students-to-live-in/
46 http://3dprint.com/107272/archinaut-nasa-made-in-space/; https://3dprint.com/110580/redworks-mars-and-beyond/
47 https://gulfbusiness.com/dubai-plans-to-3d-print-25-of-buildings-by-2030/
48 https://3dprintingindustry.com/news/montana-becomes-the-first-us-state-to-give-regulatory-approval-for-3d-printed-walls-in-construction-213806/

8 Food

Food is a necessity in life. It is, therefore, not surprising that the industry represents a high value, even though many food items are relatively cheap compared with the purchase of clothes or a house. The global food and agricultural industry was worth about $9 trillion in 2020. This is about 10% of the world's GDP[1] and is close to what is generated in the construction industry. Conventional food production and distribution promote mass standardization, leading to over-exploitation of resources, high emissions and food waste, while shipping food requires food additives to preserve the flavor and desirability (Jayaprakash et al., 2020). Agriculture accounts for about 25% of greenhouse gas emissions (Magomedov, Khaliev and Bagov, 2020). These gases, particularly CO_2, cause long-term climate change (Ullah et al., 2018). Furthermore, about one third of food produced for human consumption gets lost or wasted (Campoy-Muñoz, Cardenete and Delgado, 2017), while by 2019, globally, one in nine people in the world (815 million) was under-nourished, over 30 million children under the age of five were dangerously underweight, and, in Africa, one person in every five still goes hungry (Hasegawa et al., 2019). AM has the potential to make a positive contribution to the worldwide food challenges and can help to reduce the lack of food in developing countries (Baiano, 2020).

What is being 3D printed in the food industry

The 3D food printing market has been predicted to reach around $525 million by 2023 (Jayaprakash et al., 2020). As in the 3DP of apparel, several designers have latched onto 3DP of food. Chloé Rutzerveld, who studied industrial design, has designed a variety of foods, including with 3DP, and Daniel Wilkens, who also studied industrial design, has created a 3D printer that prints cakes. The cakes, once sliced, show different images in each slice.[2] The ability to customize (wedding) cakes has received more attention, such as by a Dutch supermarket in 2015. While customers wait, a 3D printer extrudes chocolate onto a cake in whatever shape customers desire.[3] Desserts such as tiramisu have also been 3D printed.[4] Wedding cakes often have a couple figurine at the top of the cake and this can be 3D printed in the

DOI: 10.4324/9781003399100-10

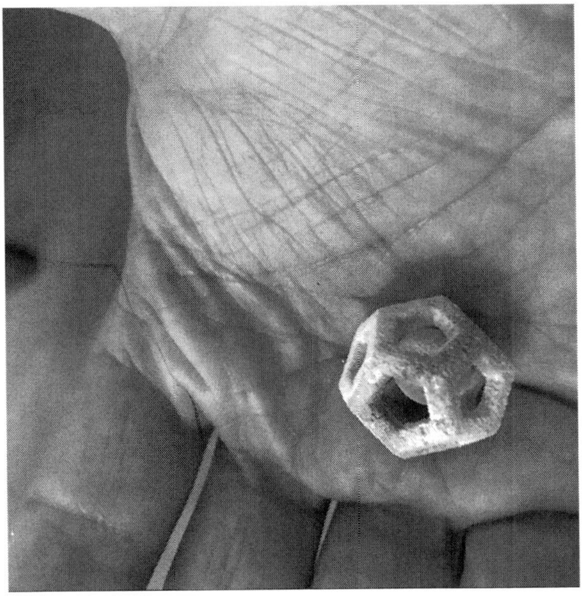

Figure 8.1 Example of 3D printed sugar made by 3D Systems.
Source: author.

likeness of the actual couple by using sugar.[5] Chocolate is comparable to concrete in terms of 3DP because at some stages it can flow easily while it can also be solid. This makes chocolate a suitable material for 3DP. Sugar is another material that works well for 3DP because of the material properties, and a variety of sugary shapes can be printed (see Figure 8.1 for an example). Sugar Labs is one of the pioneers in this field and has been 3DP sugar 'candy' since 2012.

Vegetables are 3D printed[6] as is a more complicated item such as pizza.[7] Pasta is 3D printed in a variety of shapes by BluRhapsody.[8] A perhaps less expected material that has been used for 3DP food is milk.[9] Fish is also 3D printed,[10] and in the context of an increasing realization that farming may not be efficient and that plant-based meats use less land than used for cattle farming, meat products have also been 3D printed. Some are based on bioprinting of real fat and muscle cells,[11] while others are plant-based[12]; there are even meat substitutes based on cocoa butter.[13] Food based on algae and aimed at children has been 3D printed in the shape of figurines such as Pokémon characters and animals[14] to make eating fun. Insects can provide an alternative source of protein, but especially in Western countries, consumers are resistant to eating insects. AM allows the mixing of insects with traditional foods and presenting this in a form that is more acceptable to consumers, such as demonstrated by Singapore University of Technology and Design.[15]

In addition to 3DP food, AM has been used for making food preparation items, such as 3D printed labels that come in handy when growing food,[16] and molds for use with chocolate.[17] Cookie cutters are also a type of mold, and they are available in a large variety on Thingiverse. The use of 3DP for prototyping also occurs such as for the Easispread, a heated butter knife.[18] A bit different is the use of AM for customizing coffee by printing images and messages onto coffee foam.[19] There is even a restaurant that just serves 3D printed food, and all other items such as seating and utensils are 3D printed as well.[20]

Performance: why AM printing is used for food

A key benefit of AM for food is that it provides the possibility to customize (i.e., *product flexibility*). Customization has two meanings for food. First, there is customization in terms of, roughly speaking, what it looks or feels like. AM allows food to be printed in (complex) shapes that are otherwise difficult or impossible to create (Godoi et al., 2019b). 3D printed food can alleviate chewing and swallowing issues while providing an acceptable appearance and texture (Baiano, 2020). An illustration of this is that a group of scientists is working with Unilever to 3D print chocolate that is more pleasurable to eat than traditionally produced chocolate.[21] The second type of customization is personalized nutrition (*service innovation*), a concept that has already been embraced by the US Army.[22] AM offers the possibility to tailor contents of food products to match an individual's health and activity level as well as their personal taste preferences (Jayaprakash et al., 2020).

3D printed food can have entirely novel structures and flavors (*product innovation*) that are unimaginable to be produced through traditional cooking processes (Yang, Zhang and Bhandari, 2017). AM allows the production of a large variety of ready-to-eat foods suitable to satisfy the demand of both people with specific food-related diseases (e.g., celiac disease, diabetes, hypertension, and obesity) and people with personal nutritional habits (vegetarian, vegan, etc.). It can be used to produce customized meals for elderly people, athletes, pregnant women, and children by varying nutrient concentrations, reducing or eliminating undesirable substances such as anti-nutritional factors, introducing healthy ingredients such as vitamins, fibers and phytochemicals in tradition, and making the food shape and structure pleasant (Baiano, 2020).

Another advantage is the wide range of food sources (*product innovation*) that can be used and that include both traditional and non-traditional food materials such as insects and animal by-products (Baiano, 2020). 3DP is also effective in the exploitation of agri-food residues (*process sustainability*). For example, ground potato by-products can be mixed with yam powder and hot water to obtain a printable paste usable to produce air-fried snacks (Baiano, 2020).

Table 8.1 Why AM is beneficial in the food industry

Strategic dimension	*Manufacturing capability*	*Food*
Cost	Cost efficiency	This does not appear to play a role.
	Resource efficiency	Not applicable other than reduction of food waste due to customization of portion (product flexibility) and more sustainable methods (process sustainability)
Quality	Product quality	Better food as a result of customization (product flexibility), such as better nutrition and texture or shape that can, for instance, help the elderly
Time	Delivery time	Does not seem to play a role
	Time to market	Does not seem to play a role
Flexibility	Product flexibility	Food can be customized on several dimension such as shape, texture and nutrition
Innovation	Product innovation	New types of structures and food sources (materials) are introduced as well as other food-related items such as heated butter knife and cookie cutters
	Service innovation	New services such as providing personalized nutrition
Sustainability	Product sustainability	Not applicable
	Process sustainability	Use of agri-residues and plant-based meats

Source: author.

Although AM is not yet competitive with mass-produced commodities, it has great potential for highly customized objects for small quantities. This offers a great opportunity for developing nations and small manufacturers and industries. Small manufacturers can easily and quickly adapt to demand. Small industries can simplify their supply chain by creating in-house 3DP.

Lastly, AM can also be used to produce packaging for food items, and it offers the ability to share recipes. People, of course, already share recipes in the traditional way, but with AM, an additional advantage is that owing to the automation of the process, the outcome of the recipe is identical regardless of who is involved in the preparation of the food. Similar to Table 3.3, Table 8.1 summarizes the above discussion in terms of how AM affects the strategic dimension and related manufacturing capabilities for the food industry.

How food is 3D printed: use of AM technology

The four main categories for 3DP of food are MEX, PBF (selective laser sintering or hot air sintering), BJT and MJT (Liu and Zhang, 2019; Mantihal, Kobun and Lee, 2020). There are many different types of food ingredients that can be 3D printed (Voon et al., 2019). They can be divided into three categories (Baiano, 2020):

- natively printable materials: these can be directly extruded or are in powder form
- non-printable traditional food materials such as meat, rice, vegetables and fruits: these need flow and viscosity enhancers such as starch, gelatin and agar
- alternative ingredients which include novel sources of functional and bioactive compounds such as proteins and fibers isolated from insects, algae, microorganisms, and residues from agricultural and food processes.

MEX systems have the most potential in terms of providing functional benefits to consumers (Jayaprakash et al., 2020) and have been applied to soft materials such as chocolate, dough, mashed potatoes, cheese and meat paste. However, MEX is unsuitable for making complex food structures. There are similarities in MEX of food and concrete related to material properties such as the moisture content, thermal properties, rheology and tribology (the interaction of surfaces in motion) of food material (Fan et al., 2022; Mantihal, Kobun and Lee, 2020), and buckling is also a problem for food 3DP (Liu and Zhang, 2019). Furthermore, not many foods possess the characteristic where they can solidify instantly upon extrusion (Mantihal, Kobun and Lee, 2020).

Sintering and BJT are used with 3DP sugar, such as by the already mentioned Sugar Labs. The advantage of sugar is that it can create stronger structures, and it allows a more artistic design (Singhal et al., 2020; Lee, 2021).

A special case is the bioprinting of meat-like products which use cell cultures. This is much more state-of-the-art than simple extruder technologies and much more proprietary. A complication with bioprinting is that it isn't "just" the 3DP process but may involve additional steps such as cultivating cells. Several companies are pursuing this in various ways. An example is Steakholder Foods (previously MeaTech), a company from Israel. Steakholder Foods uses real cells from animals. The cells are separated into fat and muscle cells, cultivated and then turned into bio-inks that are used in a proprietary bioprinter to create a steak structure. More details on bioprinting are provided in Chapter 9. Table 8.2 provides a summary of the above discussion.

Table 8.2 AM technologies used for food

	AM process						
	MEX	*PBF*	*VPP*	*MJT*	*BJT*	*DED*	*SHL*
Food	Most commonly used		Uncommon	Most commonly used		Uncommon	

Source: author.

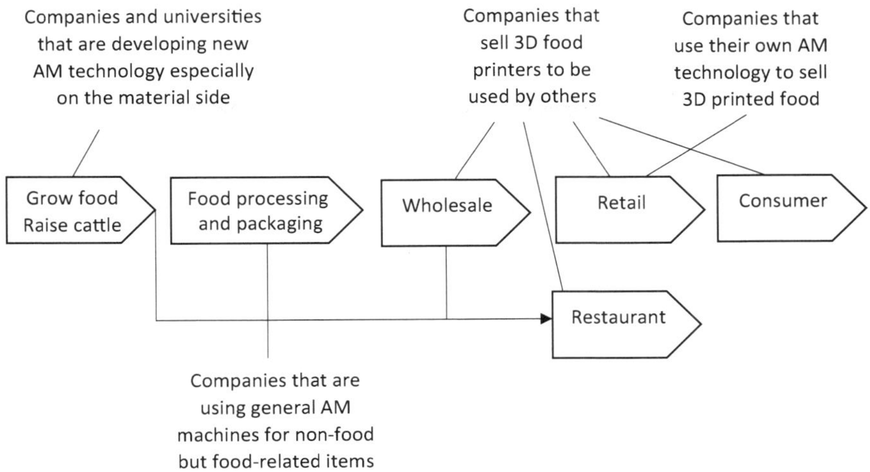

Figure 8.2 General depiction of a food supply chain.

Source: Author.

Where AM is used in the food value chain

Figure 8.2 provides a general depiction of a food supply chain. The top row of lined boxes is the general flow that ends up with a consumer who then uses this for meal preparation. The lowest lined box is to indicate the flow that ends up in a restaurant setting where the consumer then enjoys a meal.

AM affects the food supply chain in a number of different ways. At the far right of Figure 8.2 are professional food services (such as restaurants) and consumers (with their own kitchens). Some AM companies target this end of the supply chain with food 3D printers to be used by restaurants and, depending upon the price, by consumers. Businesses that use these food 3D printers can have more emphasis on the manufacturing side (make-to-stock) or more emphasis on the service side with customization (Jia et al., 2016). Examples of companies that sell these types of machines are SavorEat (the Robot Chef), byFlow (the Focus) and Natural Machines (the Foodini). Print2Taste offers professional-level machines under the brand procusini, while offering 3D printed food services at events (which can be seen as demonstrating the technology) and over time has also entered the consumer market with the mycusini brand. Some, originally more general AM, companies have introduced specific food 3D printers. An example is a food printer by Felix in The Netherlands.

There are also specialized machines such as the Pattis3 by La Pâtisserie Numérique (Digital Patisserie) for pastries, the Ripple Maker by Ripples that creates images on top of beverages such as coffee, and chocolate 3D printers such as the Choc Creator by Choc Edge and the Sweetin by Nanjing Wiiking 3D Technology.

Another approach is followed by companies such as Sugar Labs, where the in-house developed AM technology is applied to sell 3D printed food items.

To the far left in Figure 8.2 is more the ingredient side. Many (relatively new) companies are doing research on ingredients to develop better technology and food outcomes. Examples are three companies from Israel. Plantish works on 3DP plant-based salmon, Redefine Meat is working on the production of plant-based steak, and SavorEat is working on plant-based pork. Novameat is an example of a company from Spain that is working on a variety of plant-based meats. Many universities, such as Osaka University in Japan, are also involved. They are collaborating with a couple of partners to create bioprinted Wagyu beef. Another example is the Singapore University of Technology and Design, which is developing a new method of 3DP with okara (a soybean byproduct). 3DP of food is also extensively studied in the context of space travel and is funded by organizations such as NASA. BeeHex is a NASA spin-off company that became known for 3DP pizza, and NASA organized the Deep Space Food Challenge to create innovative food production technologies.[23]

Diffusion and challenges of AM in the food industry

In terms of Figure 4.6, AM has been accepted to a small degree: specialized AM machines have been made, and a few companies have adopted AM for food production purposes. However, AM is not widespread across value chains, and food that is produced with AM is not yet widely accepted. In regard to the overall adoption in rough terms, it seems that AM is not widespread and, in terms of Figure 4.7, is dealing with the very early market (i.e., the innovators). It isn't quite clear whether or how this is going to disrupt the existing markets, and much R&D in the use of AM for food products is ongoing.

The use of AM for making food continues to be faced with technical challenges such as matching food ingredients with suitable AM technologies and selection and optimization of food ingredients (Maldonado-Rosas et al., 2022). Ingredient choice for 3D food printing has a significant impact on mixing efficiency, printing speed, geometric accuracy, and compatibility with conventional post-processing techniques such as baking (Jayaprakash et al., 2020). The 3DP process is influenced by the ingredients that are printed. For example, because large size fibers can clog the nozzle, it is not simple to use MEX for fiber-rich materials.

The consumer acceptance of 3D printed food is influenced by issues such as sensory perception, knowledge about the technology, and perceived benefits (Mantihal, Kobun and Lee, 2020). One of the key issues is that 3D printed foods are perceived as not being safe. Safety is not trivial and has to do with microbial growth and requirements for printers such as how easy it is to clean Baiano, 2020; Verma et al., 2022). There is also uncertainty about regulatory (US Food and Drug Administration) approvals.

Because 3DP is in an early stage and is facing technological uncertainties, there are questions about the business model. Having a financially sustainable business can be challenging. For instance, several AM machines seem to be no longer offered in the market (Pancakebot, XYZprinting's food printer, and 3D Systems' ChefJet). The use of food 3DP is still somewhat limited to academia (research) and specialized situations that are not designed for large-scale applications, and it is not (yet) competitive in terms of economies of scale or speed. There is a lack of well-defined customer segments with profit mechanisms, and the potential of the consumer-driven business model is limited in the next few years (Jayaprakash et al., 2020).

Notes

1 https://www.plunkettresearch.com/industries/food-beverage-grocery-market-research/
2 https://3dfoodprintingconference.com/3d-food-printing-conference/message-cake-3d-printer-prints-cakes-pictures-inside-presented-daniel-wilkens-wilkens-design/
3 https://www.3ders.org/articles/20150206-dutch-supermarket-giant-albert-heijn-debuts-a-chocolate-3d-printer.html
4 https://3dprintingindustry.com/news/natural-machines-3d-prints-the-worlds-best-tiramisu-210189/
5 https://sugarlab3d.com/pages/weddings
6 https://www.3dnatives.com/en/novel-method-of-3d-printing-vegetables-110220215/
7 https://www.businessinsider.com/beehex-pizza-3d-printer-2017-3
8 https://3dprint.com/292997/chef-tim-anderson-blurhapsody-and-the-possibilities-of-3d-printed-pasta/
9 https://3dprint.com/273181/sutd-researchers-use-diw-3d-printing-and-cold-extrusion-to-fabricate-models-out-of-milk/
10 https://vegnews.com/2021/2/first-3d-printed-vegan-salmon
11 https://3dprintingindustry.com/news/meatech-breaks-new-ground-by-3d-bioprinting-a-3-67oz-steak-its-largest-yet-201005/
12 https://3dprint.com/286993/redefine-meats-plant-based-3d-printed-meat-available-in-europe/
13 https://3dprintingindustry.com/news/researchers-successful-in-3d-printing-new-cocoa-butter-based-meat-substitutes-201180/
14 https://www.3dnatives.com/en/chile-experts-develop-food-from-algae-210620224/
15 https://www.3dnatives.com/en/3d-print-foods-alternative-protein-insects-120820224/
16 https://3dprint.com/292529/3d-printed-wheat-bran-combines-with-uv-printing-to-create-colorful-designs/
17 https://www.3dnatives.com/en/hawaiian-students-3d-print-molds-for-local-chocolate-shop-270520224/#!
18 https://all3dp.com/proto-labs-easiscoop-easispread/
19 https://3dprint.com/172090/ripple-maker-nitro-cold-brew/
20 https://www.zdnet.com/article/the-restaurant-where-everything-is-3d-printed-from-the-food-to-the-chairs/
21 https://3dprintingindustry.com/news/scientists-work-with-unilever-to-develop-new-more-pleasurable-to-eat-chocolate-using-3d-printing-208134
22 https://3dprint.com/118070/us-army-3d-print-custom-meals/
23 https://www.deepspacefoodchallenge.org/

9 Medical

Broadly defined, medical here is viewed as things related to the health of the human body. Medicine, which allowed major improvement in health and life expectancy, is one of six factors that explain economic success of countries (Ferguson, 2011). Global spending on health reached $8.5 trillion in 2019 (World Health Organization, 2021). Among the challenges with healthcare are the cost and how to cover this. An example of this is the Affordable Care Act in the US, which seemingly has as many proponents as opponents. The approach of countries towards healthcare generally follows one of four basic systems: the Beveridge model, the Bismarck model, the National Health Insurance model, and the out-of-pocket model.[1] The difference between these models is how healthcare is financed. In the Beveridge model, healthcare is provided largely by government health service providers and financed through tax payments, while the out-of-pocket model is based on competition through private insurers and health service providers. AM plays a role in healthcare as it offers the possibility of improvements in effectiveness and efficiency (cost). Healthcare is also primed for AM because it often involves customization and it already has in place technologies such as CT and MRI scans that create medical images. These digital images can be used for 3DP purposes (see Rybicki and Grant, 2017).

What is being 3D printed in the medical field

A wide variety of items are 3D printed in the medical field and with many different purposes. One of the early applications is the production of prosthetics. Numerous examples exist where a high school got involved in 3DP and made a prosthetic, typically a hand, for a person in need[2, 3] (see Figure 9.1 for an example). There is even a global network of volunteers to 3D print hand, arm and upper limb assistive devices (e-Nable). There are also location-specific initiatives such as non-profit Island Missions Support for 3DP prosthetics for amputees in isolated areas of Guatemala.[4] Prostheses have been able to mimic human movements with the help of electronics[5] and have improved in strength.[6]

DOI: 10.4324/9781003399100-11

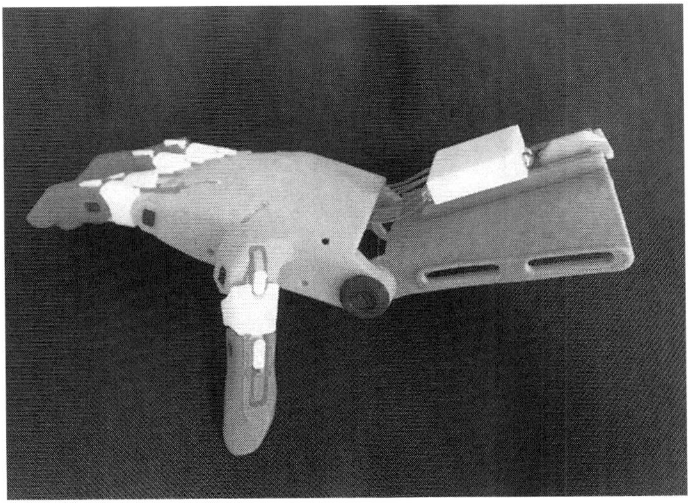

Figure 9.1 Example of a 3D printed hand prosthesis based on e-Nable design.
Source: Author.

Other types of prosthetics that have been 3D printed include legs,[7] jaw,[8] eye[9] (and, related to this, lenses for glasses,[10] frames for glasses,[11] sunglasses,[12] contact lenses[13] and customized tinted lenses that help with color blindness[14]), an ear,[15] and faces for people with face deformities, for instance, due to medical procedures.[16] 3DP also found an application in allowing replicas of a person's face, in case of deformities from the cause of death, so that open caskets can be used for funerals.[17] On that topic, custom cremation urns have been 3D printed in the shape of the deceased's head.[18] Prosthetics have also been 3D printed for animals, allowing a dog with deformities to run[19] and improving the lives of other animals such as a duck, penguin, lizard, cat and turtle.[20]

Hearing aids constitute another major area where AM has been applied. Almost 100% of the custom hearing aid market is 3D printed today.[21] Similarly, broad applications exist in the dental field with 3D printed dental aligners[22] and crowns.[23] Some examples provided in Figure 9.2 were created by 3D Systems and presented at 2022 FormNext. Furthermore, a variety of medical implants have been made with AM, such as knees[24] (including bioprinting of cartilage tissue[25]), chest bone and ribs,[26] and there are trials for bioprinted breast implants.[27]

3DP has also been used for living cells[28] and tissue which extends into 3DP of organs (bioprinted organoids with demonstrated key liver functions[29]) and heart valves.[30] In fact, a beating human heart has been printed by several groups of researchers.[31] In a slightly different context, models of organs have been printed to serve as a tool to plan for surgery[32] and as "replacement" organs for deceased individuals.[33]

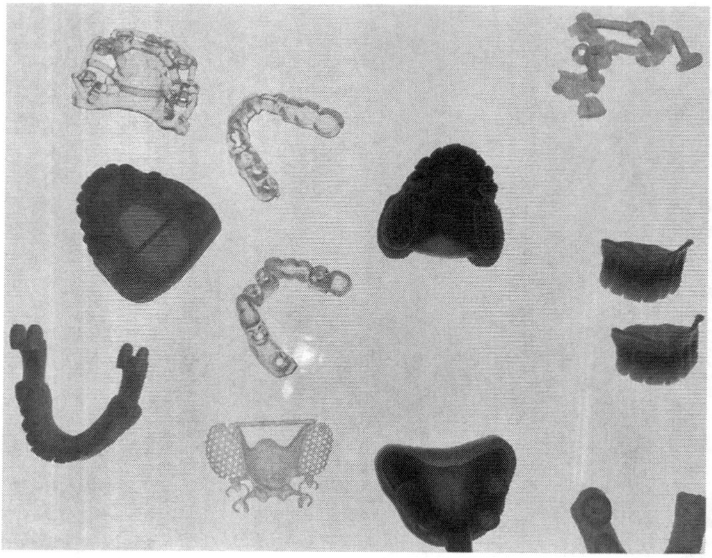

Figure 9.2 Examples of dental AM applications.
Source: author.

AM is also pursued in the area of pharmaceuticals because of the ability to customize medication that releases drug doses at different body regions,[34] and a smartphone-based 3D printer has been used to print personalized medicine.[35] During Covid, AM was used to produce face shields,[36] respirators[37] and Covid-19 tests.[38] Even in forensic science, AM has found applications and has helped to solve crimes.[39] Similar to construction and food, medical AM is also of interest for space travel. 3D bioprinted skin bandages have already been tested in space,[40] and work is ongoing on a variety of other AM applications in space.[41]

Completely new applications have also been created because of AM, such as the biopen (a handheld device for surgical printing; O'Connell et al., 2016), wearable biosensors for sunburn protection[42] or help with medical issues,[43] and 3DP of ultrasound images so new parents can have a 3D model of the fetus.[44] Examples of improved products that have become possible with AM are a toothbrush for people with dexterity impairments[45] and a surgical fracture table which cut cost by 98%.[46]

Performance: why AM is used in the medical field

One reason for the use of AM is cost advantage (*cost efficiency*). This is particularly true for the manufacturing of simple prostheses. With traditional methods, a prosthesis may cost up to $50,000, but with AM this can be done for $10 or less.[47]

Customization (*product flexibility*) in combination with cost is another major reason for the use of AM in the medical field. People are different and thus there is a very high need for customization (and low volume), but this has been too costly to accomplish with TM (Chen et al., 2020b; Sandström, 2016).

Reduction of time, in combination with the above, is another reason to use AM (*delivery time*). For hearing aids, the traditional method of creating a hearing aid involved creating a silicone model of the patient's ear canal, having highly skilled technicians create the hearing aid shell through a series of manual steps, including sculpting, molding and curing, and then integrating the electronic components into the shell. AM removes the time-consuming manual labor involved in creating the shell. Work that previously took one hour could in some cases be done in five minutes, and for some companies half the number of people are now making four times as many shells per unit of time (Sandström, 2016). AM also provides opportunities to produce closer to the hospital site. For instance, traditional biomedical implants are typically available from suppliers that are out of the region. AM has the potential to provide patient-specific customized parts with a faster response, a lower inventory level, and reduced delivery cost (Emelogu et al., 2016). Developing nations provide another example where medical resources may be limited but local AM facilities can help, such as at the Sierra Leone project.

Quality, in terms of *process consistency*, was discussed in Chapter 3, where it was explained that process consistency is typically a disadvantage for AM. An exception to this is the medical field. Making hearing aids with TM involved manual labor that was based on skilled technicians whose work is largely an art. This was difficult to control and reliability was low (Sandström, 2016). AM allows more standardization of the production process, which improves the overall quality.

AM can also enhance quality (*product quality*) because of the ability to create complex shapes. Custom implants can possess complex features that are difficult to make with TM (Emelogu et al., 2016). In the case of pharmaceuticals, it allows for personalized doses (Öblom et al., 2019) and the polypill (a single tablet that includes the combination of several drugs) (Jose and Christopher, 2018).

3D printed organs can be viewed as improving *process sustainability* as 3D printed organs reduce the need for donors. For instance, cardiovascular diseases are the number one cause of death in industrialized nations, but the number of cardiac donors is limited. Using AM provides an alternative approach which furthermore provides the advantage that the organs are personalized, which can reduce or eliminate the risk of implant rejection (Noor et al., 2019) (i.e., *product sustainability*). Work is also progressing on bioprinting, which reduces the need for animal testing,[48] which is another illustration of *process sustainability*. Examples of *product innovations* are new surgical tools that are possible because of AM, such as the biopen (O'Connell et al., 2016) and Sutrue's automated suture stitching devices[49] (which can

Table 9.1 Why AM is beneficial in the medical field

Strategic dimension	*Manufacturing capability*	*Medical*
Cost	Cost efficiency	Several items, such as prosthesis, can be produced cheaper.
	Resource efficiency	Not applicable other than what is already covered on, for example, improvements due to product and service innovation
Quality	Product quality	Customized products, such as implants or prosthesis, improve performance.
Time	Delivery time	Customized parts, such as hearing aids, can be delivered faster than with alternative manufacturing methods.
	Time to market	Rapid prototyping allows new products such as surgical tools to be brought quickly to the market.
Flexibility	Product flexibility	Customization allows tailoring to customers. such as with implants and medicine.
Innovation	Product innovation	Many new products/applications have been introduced in the medical field.
	Service innovation	Customization has allowed service innovation such as 3DP of anatomical models of the patient which improves the service of medical operations.
Sustainability	Product sustainability	New materials such as with bioprinting
	Process sustainability	3D printed organs are a more sustainable method to ensuring availability of organs than reliance on donors.

Source: author.

also be brought to the market relatively fast, *time to market*), and the production of patient-specific anatomical models which can reduce risk to the patient by decreasing surgery time, reducing the probability of the surgical team encountering unexpected anatomy or relative positioning of structures and/or devices, and better pre-operative planning of the surgical workflow (Parr et al., 2019) (i.e., *service innovation* and related to it process quality). Similar to Table 3.3, Table 9.1 summarizes the above discussion in terms of how AM affects the strategic dimension and related manufacturing capabilities for the medical field in particular.

AM methods that are used for medical applications

Owing to the large variety in AM applications in the medical field, there is also variety in AM technologies that are applied. Three of the main AM technologies are frequently applied, two less frequently and two rarely (Salmi, 2021).

Frequently applied are MEX (such as for simple plastic prosthesis and models [Ravi et al., 2022]), PBF (such as for medical devices including [metal] implants [Novakov et al., 2017] and surgical tools [Mitsouras and Liacouras,

2017]), and VPP (such as for hearing aid shell [Sandström, 2016] and dental [Santos et al., 2022]). This includes variants such as Carbon's use of CLIP (a variant of VPP) for Covid nasal swab tests.[50] Less frequently applied are BJT (applied for pharmaceutical pills [Chen et al., 2020b]) and MJT (applied for anatomical models [Tyagi, Yadav and Deshmukh, 2022]).

A special case is the use of bioprinting. Bioprinting is not a separate type of AM process but has to do with what is printed; that means it involves organic material such as tissue and cells. Since it may involve live cells, some of the challenges are (a) to 3D print the cells while keeping them alive during the 3DP process, (b) to have systems in place that allow the cells to remain alive after the 3DP process, and (c) to have structures in place so that subsequent layers are supported and the structure doesn't collapse. One approach has been to use cell-supportive hydrogels which protect cells during the 3DP process, have material properties (stiffness) to retain its shape, and include proteins that support cell growth after the 3DP process. Bioprinting is especially applied in regenerative medicine to address the need for tissues and organs suitable for transplantation and takes place using a variety of AM processes such as MEX and MJT (Mandrycky et al., 2016; Skardal, 2018). A variety of tissues have been created with bioprinting. Examples include bone tissue, cardiac tissue, heart valves, liver tissue, lung tissue, skin tissue, and vascular tissue (Ozbolat, Peng and Ozbolat, 2016). Three central approaches in bioprinting are biomimicry, autonomous self-assembly, and mini-tissue building blocks (Murphy and Atala, 2014). With biomimicry, each individual component of native tissue is engineered. The challenge is that it is often extremely difficult to reproduce all elements that make up the milieu of a given target tissue (Bishop et al., 2017). Autonomous self-assembly uses embryonic organ development as a guide and relies on factors like structural and functional properties whereby cells develop into tissue (Ashraf et al., 2018; Murphy and Atala, 2014). The characteristic of a typical complex *in vivo* tissue, which is composed of many simpler units whose combination contributes to the whole, is what microtissues rely on. In this strategy, tissue engineering is used to form the smallest structural and functional units which can be combined into the final target tissue (Murphy and Atala, 2014). It goes beyond the scope of this book to provide a more detailed discussion of the diverse field of bioprinting, but good insight can be gained from Gu et al. (2020) and Khademhosseini and Camci-Unal (2018). Table 9.2 provides a summary of the above discussion.

Table 9.2 AM technologies used in the medical field

	AM process						
	MEX	*PBF*	*VPP*	*MJT*	*BJT*	*DED*	*SHL*
Medical	Most commonly used			Less prevalent but used in bioprinting	Less prevalent	Uncommon	

Source: author.

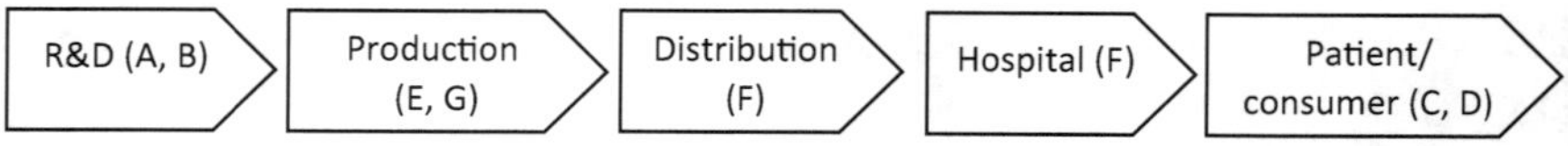

Figure 9.3 Medical value chain.
Source: author.

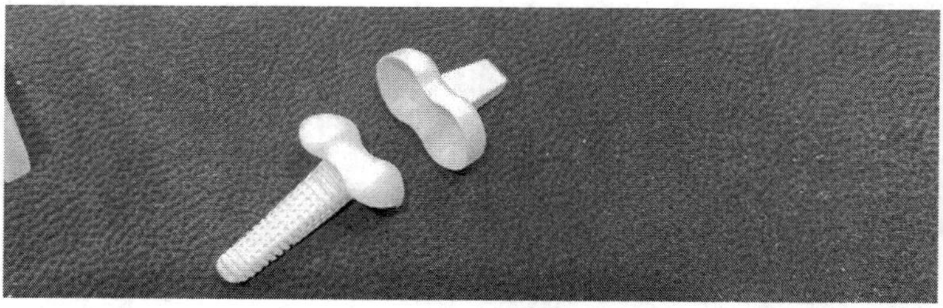

Figure 9.4 3D printed finger joint.
Source: author.

Where in the medical value chain

The value chain of the medical field is complex. While examples exist that depict a medical supply chain (see Kim and Kim, 2019), this depiction depends upon the 'product' involved. Therefore, a very broad and general chain is provided in Figure 9.3 for illustrative purposes and to facilitate the below discussion.

AM affects the medical value chain in a variety of ways due to the variety of AM applications in the medical field. In some instances, AM replaces TM; in others, it expands upon TM. There are also situations where it has created completely new businesses and applications. The AM impact on different parts of the medical industry value chain involves at least the following situations:

A. On the left-hand side, research is conducted on the application of AM, such as for organs. This research involves the use of AM to make (medical) progress towards creating fully functional organs. Research takes place at universities in combination with hospitals such as the Wake Forest Institute for Regenerative Medicine and the Lewis Lab at Harvard University. Research is also conducted at companies such as Organovo. It can also involve partnerships such as joint research by Finland-based Brinter employees and researchers at the Tampere University.[51]
B. A slight variation on this theme is the development and production of the equipment and material, including bioprinters and materials for bioprinting. Sometimes, there is overlap because working on improved cells and tissues may lead to required adjustments in materials for bioprinting and/or bioprinting processes (i.e., AM equipment). Organovo is an example of a medical company that is developing human tissue and in order to do

this developed their own bioprinter. Examples of companies that work on bioprinters and/or materials are the BICO group (previously CELLINK, from Sweden), Allevi (from the US and acquired by 3D Systems in 2021), Inventia (from Australia), Advanced Solutions Life Sciences (from the US) and Rokit Healthcare (from South Korea). Figure 9.2 is an example of a 3D printed ceramic finger joint that was made with a Lithoz VPP process and displayed at the 2022 FormNext tradeshow.

C. A variation on this theme is that in the consumer space there are specialized desktop 3D printers that can be adjusted for bioprinting purposes, which provides laboratories with cheaper bioprinting options but this also potentially leads to consumer-level bioprinting activities. Examples are Prusa[52] and Aleph Objects.[53]

D. On the right-hand side, on some occasions consumers get involved with 3DP complicated things such as braces,[54] but mostly it involves applications such as manufacturing simple prosthesis such as the e-Nable designs.

E. On the production side, there are companies that utilize 3D printers. An example of this is the hearing aid industry, where customization was already part of the business model and where 3D printers provide a great fit. Moving more into this space are pharmaceutical companies with personalized medicine.

F. Connected to this are distribution which gets into hospitals and the redistribution of manufacturing in healthcare.[55] AM provides opportunities to produce closer to the point of use, which in many cases is the hospital. Hospitals have been developing AM capabilities, and more than 160 hospitals in the US operate AM facilities (Wohlers Associates, 2020). These can be divided into different types of capabilities. Some capabilities relate to bioprinting labs (Sanicola et al., 2020). This connects to R&D, as discussed under item A. Other capabilities relate to 3DP facilities that 3D print anatomical models that can be used to prepare for surgery (Sheikh et al., 2017). An example is the 3D innovations lab at the Rady Children's Hospital in San Diego. However, whether such decentralized facilities in hospitals are cost-effective depends on the type of AM and materials used (Verboeket et al., 2021).

Specialized (centralized) outside parties in AM also offer medical AM services, such as pre-operative models. Examples are Oceanz and 3D LifePrints. Dentistry provides another example where centralized scanning and manufacturing facilities have been replaced by decentralized laboratories (Dawood et al., 2015). In this regard, owing to the trends in online shopping, many shopping centers and malls have experienced lower occupancy rates, but the medical and dental industries have been expanding into the open spaces; this is partially due to growth that is enhanced with AM allowing better service.[56]

G. There are also completely new business or products due to the application of AM that may or may not get distributed or used in hospitals, such as the previously mentioned 3D printed version of an ultrasound, face prothesis to aid open-casket funerals, and newly developed tools for surgical procedures such as Sutrue's stitching device.

H. Changes extend into education as well. Education isn't shown separately in Figure 9.3 but influences a value chain because it influences the skillsets of people across the value chain. Several AM applications in the medical field require bio/medical knowledge as well as design/manufacturing knowledge. This has given rise to completely new educational programs combining both, such as for biofabricator. The University of Utrecht in The Netherlands,[57] the University of Bayreuth in Germany,[58] Newcastle University in the UK,[59] and the University of Wollongong in Australia[60] all offer a biofabrication degree program at the master's level.

Diffusion and challenges of medical AM

In terms of Figure 4.6, the concept of AM has been accepted, AM machines have been adopted, and products that have been produced with AM, such as prothesis and anatomical models, have been accepted. Some areas of AM, such as for hearing aids and dental applications, are very well established. These may already be towards the end of the early majority or even in the late majority stage of adoption. Other areas are at a very early stage such as prosumer-printed frames for glasses, while in some fields, AM is only in an early phase of research, such as 3DP organs. In regard to the overall adoption in rough terms, it appears that despite the described instances of AM adoption and the few very-well established AM adoption areas, 3DP is not widespread considering the scale of healthcare. Instead, AM in healthcare appears to be, in terms of Figure 4.7, at an early stage where technology enthusiasts (innovators) and visionaries (early adopters) have been adopting AM and the chasm has not been crossed.

A main challenge for further diffusion in medical AM, as in construction and food, is concern for safety, which is linked to regulation. Certain aspects of the medical industry are well regulated, but others are not (Chen et al., 2020b; Mitsouras et al., 2017). For example, during Covid-19, many well-intended owners of desktop 3D printers got involved in producing items to help fight Covid. The safety and reliability of the parts often lacked approval (Singh, Venkatesh and Deoghare, 2021). Unfortunately, the produced items didn't always perform. Bezek et al. (2021) showed through testing of 3D printed respirators that almost all provided less than 60% filtration efficiency, significantly below the required 95% efficiency of the N95 respirator, thus providing a false sense of protection. As with the construction industry, getting approval from regulators is expensive, difficult and time-consuming when it comes to new technology[61]; for medical applications, clinical trials have to be approved.[62]

Additional challenges with AM are that the manufacturing speed is low and that there are limitations to the materials (Chen et al., 2020b). An exception is that the adoption in the hearing aid industry has been high. This is because of the high degree of customization so that AM provides considerable advantages compared to TM (Sandström, 2016). Products that require higher and perhaps more standardized quantities may (still) benefit from TM approaches. The availability of materials plays a role in pharmaceutical AM (Chen et al., 2020b) and bioprinting (Mandrycky et al., 2016) and for medical implants which tend to use solid 3D print materials such as titanium, which causes new stresses on nearby bones (Garner, Wu and Zadpoor, 2022).

For bioprinting, there are additional complexities in terms of the reliability and technical challenges related to the sensitivities of living cells and the construction of tissues (Murphy and Atala, 2014). Bioprinting usually involves production of small-scale tissues for use in labs, but even if that is successful, they are not easily scalable to full size (Sun et al., 2020). An example of the challenges is the kidney. It is not yet possible to model the architectural complexity of a kidney, so it still is unavailable for lab situations, let alone the ability to 3D print a functioning kidney (Fransen et al., 2021). A kidney is quite different from, for example, a heart that contains a lot of muscle tissue, which is easier to 3D-print (Yadid et al., 2022).

Notes

1 http://www.pbs.org/wgbh/pages/frontline/sickaroundtheworld/countries/models.html
2 https://www.businessinsider.com/kylie-wicker-boylan-high-school-3d-printed-hand-2014-4
3 https://nbc16.com/news/local/high-school-students-print-prosthetic-hand-for-boy
4 https://3dprintingindustry.com/news/slice-engineering-technology-helps-island-mission-support-3d-print-prosthetics-for-remote-amputees-213698/
5 https://3dprint.com/293235/3d-printed-robotic-hand-mimics-human-movement
6 https://www.3dnatives.com/en/robotic-hand-lift-1000x-its-weight-100820224
7 https://www.3ders.org/articles/20141224-3d-printed-exo-prosthetic-leg-balances-aesthetics-with-affordability.html
8 https://www.3dnatives.com/en/first-3d-printed-titanium-jaw-successfully-transplanted-on-cancer-patient
9 https://3dprintingindustry.com/news/london-man-receives-worlds-first-3d-printed-prosthetic-eye-200197
10 https://3dprintingindustry.com/news/breakthrough-coating-technology-makes-3d-printed-lenses-anti-reflective-208682
11 https://www.sculpteo.com/en/3d-learning-hub/applications-of-3d-printing/3d-printed-glasses/
12 https://all3dp.com/2/best-3d-printed-sunglasses/
13 https://www.fabbaloo.com/news/3d-printing-and-ar-contact-lenses
14 https://3dprint.com/293973/researchers-3d-printing-customized-tinted-lenses-to-correct-color-blindness/
15 https://www.fabbaloo.com/news/3d-printed-ears-first-used-in-human-patient
16 https://www.medicalopedia.org/4162/3d-printed-face-transplant-changes-a-mans-life/
17 https://3dprint.com/169038/3d-printing-chinese-funeral-home/
18 https://www.fabbaloo.com/news/3d-printing-cremation-urns?
19 https://www.youtube.com/watch?v=uRmoowIN8aY

20 https://www.sculpteo.com/blog/2018/05/02/3d-printed-prosthetics-8-incredible-animal-prostheses/
21 https://www.3dsourced.com/editors-picks/custom-hearing-aids-3d-printed/
22 https://3dprintingindustry.com/news/prodways-machines-to-3d-print-up-to-a-million-dental-aligners-per-year-in-largest-ever-project-200727 and https://3dprintingindustry.com/news/dental-3d-printing-stock-soars-63-after-walmart-deal-and-lifting-of-trading-restrictions-167347/
23 https://www.dentistrytoday.com/2021-the-year-of-3d-printing-in-dentistry/
24 https://www.thefabricator.com/additivereport/article/additive/how-3d-printing-and-robots-improve-knee-replacement-surgery
25 https://www.fabbaloo.com/news/bioprinting-cartilage-tissue-to-help-knees-function-again
26 https://3dprintingindustry.com/news/3d-printed-titanium-chest-implant-saves-cancer-patients-life-203854
27 https://3dprintingindustry.com/news/collplant-to-trial-3d-bioprinted-breast-implants-in-breakthrough-animal-study-207127 and https://www.3dnatives.com/en/3d-printed-scaffold-breast-implants-090820225/
28 https://3dprintingindustry.com/news/matricelf-moves-closer-to-curing-paralysis-with-in-house-3d-printable-human-stem-cells-208730/
29 https://3dprint.com/289069/volumetric-bioprinter-3d-prints-liver-organoids-in-less-than-20-seconds/
30 https://3dprintingindustry.com/news/researchers-3d-print-regenerative-heart-valves-that-grow-with-patients-206122/
31 https://3dprintingindustry.com/news/beating-heart-3d-printed-by-cas-scientists-on-modified-bioprinter-204887/ and https://3dprintingindustry.com/news/boston-universitys-new-3d-printed-mini-human-heart-beats-on-its-own-208314/
32 https://www.3dnatives.com/en/3d-printed-human-heart-model-251120205/ and https://3dprint.com/2282/3d-print-cancerous-kidney-model/
33 Because the removal of organs (due to organ donation) compromises the shape of the body which is undesirable in certain cultures, see https://www.fabbaloo.com/news/3d-printed-organs-for-the-dead
34 https://3dprintingindustry.com/news/3d-printing-breakthrough-enables-custom-tablets-to-be-produced-in-just-seven-seconds-206835/
35 https://3dprint.com/287752/researchers-develop-smartphone-based-3d-printer-for-fabricating-personalized-medication/
36 https://www.theverge.com/2020/5/25/21264243/face-shields-diy-ppe-3d-printing-coronavirus-covid-maker-response
37 https://www.simplify3d.com/lessons-from-the-field-3d-printing-respirators/
38 https://3dprintingindustry.com/news/carnegie-mellon-university-develops-10-second-covid-19-test-using-optomec-aerosol-jet-process-176330/ and https://www.carbon3d.com/news/press-releases/resolution-medical-3d-prints-swabs-for-covid19
39 https://www.additive-x.com/blog/how-3d-printing-helped-solve-the-murder-case-of-ellie-butler/
40 https://3dprintingindustry.com/news/astronauts-test-3d-bioprinted-skin-bandages-in-space-202445/
41 https://3dprint.com/274114/bioprinting-in-microgravity-where-do-we-stand/ and https://3dprint.com/287740/handheld-bioprinter-could-one-day-heal-astronaut-injuries/ and https://3dprint.com/294766/iss-looks-for-experiments-for-tissue-engineering-and-bioprinting-in-orbit/ and https://3dprintingindustry.com/news/redwire-to-launch-improved-3d-bioprinter-to-the-iss-on-mission-to-aid-us-military-veterans-216487/
42 https://3dprintingindustry.com/news/clarkson-university-scientists-develop-3d-printed-wearable-biosensors-for-sun-burn-protection-179750/

43 https://3dprint.com/294860/3d-printed-sensors-developed-to-aid-lupus-patients/
44 https://all3dp.com/expectant-parents-can-now-get-a-3d-printed-model-of-their-unborn-child/
45 https://3dprintingindustry.com/news/dentists-develop-unique-2-in-1-3d-printed-toothbrush-handle-for-people-with-dexterity-impairments-212767/
46 https://3dprint.com/292824/researchers-develop-low-cost-open-source-3d-printed-surgical-table/
47 Owing to variety, costs diverge a lot for example https://www.businessinsider.com/kylie-wicker-boylan-high-school-3d-printed-hand-2014-4 mentions up to $50,0000 and https://3dprintingindustry.com/news/scientists-create-fully-automated-3d-printed-prosthetic-production-line-185892/ states $3,000-$100,000 for a hand while https://3dprint.com/263051/a-3d-printed-prosthesis-helps-a-brazilian-mother-recover-her-missing-piece/ mentions $500,000 for a face prosthesis.
48 https://3dprintingindustry.com/news/tu-wien-3d-prints-human-tissue-on-a-chip-in-bid-to-eliminate-animal-testing-210599/
49 https://3dprint.com/216024/3d-printed-suture-stitching-device/
50 https://www.carbon3d.com/news/press-releases/resolution-medical-3d-prints-swabs-for-covid19
51 https://www.brinter.com/press-and-news/brinter-team-member-contributes-to-scientific-bioink-research/
52 https://3dprint.com/248126/nydus-one-syringe-extruder-nose-evaluating-open-source-bioprinting-modifications-prusa-i3/
53 https://3dprintingindustry.com/news/aleph-objects-enters-3d-bioprinting-with-fluidform-hardware-coming-summer-2019-156914/
54 https://www.sciencealert.com/a-college-student-has-3d-printed-his-own-braces-for-less-than-60
55 Note: a consortium of seven UK universities and the MoD Defence Medical Services seeks to identify, fund and advance high-value applications of redistributed manufacturing research that will transform future deployed medical care capability. See for their interesting results: www.rihn.org.uk.
56 https://www.fabbaloo.com/news/3d-printing-and-the-impact-of-new-medical-retail-spaces
57 https://www.uu.nl/masters/en/biofabrication
58 https://www.uni-bayreuth.de/en/master/biofabrication
59 https://www.ncl.ac.uk/postgraduate/degrees/4864f/
60 https://www.uow.edu.au/research-and-innovation/our-research/research-institutes-and-facilities/australian-institute-for-innovative-materials/masters/
61 https://3dprint.com/294813/ossiform-raises-6-7m-for-3d-printed-bone-implants/
62 https://www.3dnatives.com/en/triastek-inc-t21-fda-approval-06122022/#!

10 Transport

Transport is considered broadly in this chapter and includes transport by road, water and air. The car industry, owing to its employment, is often considered important politically, as evidenced by the 2008 financial crises bailout of General Motors (GM) and Chrysler. The International Organization of Motor Vehicle Manufacturers reported that, in 2021, roughly 56–57 million passenger cars were produced and sold worldwide. The car manufacturing industry is often a focus for research (see Beer, 2018; Demlehner et al., 2021), and the findings frequently have implications for other industries. Lean manufacturing, discovered in a study on Toyota, serves as an illustration (Womack et al., 1990). North America, Europe and Asia are major car manufacturing regions (Dicken, 2011), while the manufacturing of parts is spread across a worldwide network of suppliers (Abele et al., 2008).

Commercial aviation shares some of these characteristics, such as the global spread and political importance, although the number of major aircraft manufacturers is less than the number of major car manufacturers (Eriksson and Steenhuis, 2016). Shipbuilding is similar except that Asia (i.e., South Korea, China and Japan) is the main region for the production of ships (Huang and Park, 2018).

Some of the main concerns for these industries are to improve the efficiency of the production process and reduce the environmental impact. To illustrate: aviation generates about 9% of global greenhouse gas emissions (Gisario et al., 2019), and the International Civil Aviation Organization has a target of 50% reduced aviation emissions by 2050 (Blakey-Milner et al., 2021).

What is being 3D printed related to transport

For many people, "seeing is believing," and at the Detroit auto show in January 2015, an appealing event occurred. Local Motors, a company from the US, 3D printed a car in 44 hours live at the show.[1] The 3DP of cars, similar to 3DP of a house involves only certain aspects. For the Local Motors car at the show, the Strati, AM was used for the frame, exterior body and interior features. The mechanical parts such as the battery, motors, wiring, suspension

DOI: 10.4324/9781003399100-12

and tires were not printed.[2] Local Motors developed the Strati further into a smoother looking LM3D Swim and also used AM for part of the production of an off-road vehicle, the Rally Fighter. The R&D for the Rally Fighter involved community engagement, and customers were involved in the production process that took place in mobile micro-factories (Anderson, 2012).[3] In 2016, Local Motors 3D printed an autonomous minibus, the Olli, and an improved version in 2019. The Olli was tested and used in locations such as Berlin[4] and Sacramento State University.[5]

The Strati received a lot of press but was not the first nor the only 3D printed car. Several other cars were 3D printed: the Urbee,[6] the 4ekolka,[7] the Shuya[8] and the Micro Commuter by Honda.[9] These cars mostly got stuck in the prototype stage and did not go into full-scale production. AM was also used for making the lightweight chassis of a supercar, the Blade by Divergent Microfactories.[10] Divergent has also been involved in producing parts for other car manufacturers such as Aston Martin.[11] AM is used for the production not only of "entire" cars but also of parts for current,[12] classic[13] and custom cars.[14] Research has highlighted the role of nostalgia (Pascal et al., 2002), and there may be rising demand for the reproduction of classic cars. The Low Volume Motor Vehicle Manufacturer's Act, which went into effect in 2021, makes it easier for manufacturers to economically produce cars in small quantities.[15] Figure 10.1, taken at the 2018 FormNext show, gives an idea of what is possible.

AM parts include pistons,[16] obsolete parts such as the production of an original but discontinued carburetor,[17] and accessories (such as a sunglass holder, a can holder, and a phone or cardholder explicitly created for the new Peugeot 308[18]), and Ford has released CAD[19] files for owners of the 2022

Figure 10.1 AM used for cars.

Source: author.

Ford Maverick to 3D print customized components,[20] while Mercedes-Benz allows transport companies with Daimler Buses to 3D print spare parts.[21] Daimler, through its service brand Omniplus, has also created a mobile 3DP center for the decentralized production of spare parts. This is a container that can be transported to different locations so that it can produce close to where demand is.[22] Race cars constitute another area where AM has been applied as it can help with developing lighter and more aerodynamic parts which can make a car faster.[23] An example is provided in Figure 10.2.

In NASCAR, a high-performance windshield air cockpit ventilation unit was 3D printed.[24] Furthermore, as much as 70% of Alpine's Formula 1 car's bodywork parts are prototyped with AM,[25] while Alfa Romeo, another F1 team, uses AM for structural parts.[26] Supercar manufacturers such as Lamborghini have also been making parts with AM, such as fuel caps and air duct clip components.[27] There are also AM-produced parts for trucks, such as aluminum alloy thermostat covers for truck engines.[28]

AM is also used for prototyping such as the development of a light-weight car seat[29] and entirely redesigned parts such as Michelin's airless tires,[30] and another application is the use of AM to produce production tools. An example is a tool for the correct placement of logos used by Volkswagen.[31] A different type of application is that of 3D printed gloves for car assembly line workers to prevent musculoskeletal disorders.[32] Other road vehicles that AM has been used for include motorcycles[33] and bicycles,[34] including optimized bicycle saddles.[35]

Figure 10.2 Example of an AM-produced part designed by Marco Mattia Cristofori presented by BigRep at the 2018 FormNext.

Source: author.

AM has found broad applications in aviation as well; however, aviation has extensive safety regulations (Eriksson and Steenhuis, 2016; Russell et al., 2019), so the application of AM is less straightforward, and in contrast to cars, "entire" airplanes have not been produced with AM. Structural parts are the most challenging because safety approvals are needed when production methods are changed. Nevertheless, a first load-bearing metal part was approved by the European Aviation Safety Agency (EASA) in 2022 for use on airplanes. This is a small part that more or less fixes a duct in place.[36] Non-structural cabin parts such as a clip to hold back aircraft curtains and hand sanitizer and soap bottle holders,[37] luggage bins bulkheads and panels[38] are less regulated and thus easier for AM applications. In 2021, Boeing reported more than 100,000 AM parts in flight; most of these were non-metal. The Airbus A350 features over 1,000 3D-printed parts, including cabin parts, titanium pylon brackets, and a cabin spacer,[39] and the Boom Supersonic XB-1 prototype contains 21 AM parts, including ducts and various louvers.[40] One of the most publicized AM parts in aviation is a metal fuel nozzle for the CFM LEAP engine produced by GE, of which over 100,000 have been used.[41] Another example is the GE9X engine, which contains 3D printed titanium turbine blades.[42] Small jet engines have also been 3D printed.[43] Furthermore, unmanned aerial vehicles and drones are increasingly used for a variety of applications, including agriculture, emergency and disaster management, film making and security, and are also made with the use of AM.[44]

An extension of aviation is aerospace. NASA has already been mentioned in Chapters 7 and 8 and, obviously, is important for aerospace. In May 2022, NASA revealed that it is funding several projects to enhance space exploration. Among them are 24 that are either creating new AM processes or using AM technologies.[45] Private companies are also increasingly involved in the development of small spacecraft and launch vehicles for low earth orbit, particularly for satellites and communications. There are at least 50 companies involved, of which 72% are based in the US.[46] Companies such as Boeing, Fleet Space, Mitsubishi, and Privateer Space use AM to produce parts of satellites.[47] Satellites are brought into space by launch vehicles, namely rockets. Relativity Space 3D printed a complete rocket, that successfully launched in March 2023 although it did not reach orbit[48] and rocket engines have been 3D printed as well. SLM Solutions printed a liquid oxygen and propane engine in one piece,[49] Skyrora 3D printed a rocket engine,[50] and in June 2022 Rocket Lab from New Zealand launched its Electron rocket with the first oxygen/hydrocarbon (Rutherford) engine that used AM for all of its primary components. Rocket Lab was able to make lighter parts and reduce the build time from months to just days by using AM instead of TM.[51] Rocket engine parts such as the thrust chamber by Launcher and several parts by Ura Major have also been 3D printed.[52] Seats in the Boeing Starliner spacecraft have been 3D printed,[53] and AM technology has also allowed the introduction of completely new products such as a coffee cup for a weightless environment.[54] Furthermore, 3D printers have been on the International Space Station (ISS) for experimenting with AM in a zero-gravity environment,[55] and Made in Space 3D printed a ceramic single-piece turbine bladed disk on board the ISS.[56]

AM has also been used for water vehicles such as a complete kayak,[57] a small boat 3D printed with recycled plastic from milk jugs,[58] and an underwater "scooter" that is 75% 3D printed.[59] Similar to cars, boat parts are made with AM,[60] as are sections of boats[61] and also the structure of a 25-foot boat.[62] As in aviation, there are intense standards and requirements (due to corrosion, fatigue and shock resistance), nevertheless AM has been used to make parts such as for autonomous underwater vehicles[63] and functional parts such as a metal propeller spanning 2.5 meters.[64] AM is also applied on ships so that parts can be produced when the ship is at sea.[65] This may be particularly useful in remote locations such as an arctic icebreaker for which repairs and replacement of parts are challenging but AM provides a solution.[66] Some last examples come from rail, where AM is increasingly used for spare parts such as aluminum brake parts[67] and a variety of other items, including seat prototypes and complex and difficult-to-reproduce dashboard frames.[68] France-based Alstom has used MEX for the production of metal doorstoppers for trains.[69]

What this means is that, in the transportation sector, AM has been applied in a large variety of settings such as prototyping, tooling and production. For production, a distinction can be made between using AM for the production of an existing design such as spare parts and the production of a new design. Some production is "unique", meaning that it is essentially one part that is manufactured, such as a part for a unique rocket. This is different from prototyping where the part is not for actual usage. Most production is in low volumes, although there is a movement upwards towards more moderate volume levels. This is depicted in Figure 10.3. What is not shown in Figure 10.3 is the distinction between parts and a final product such as a whole car, nor the distinction between structural parts, such as a load bearing part, versus non-structural parts.

Performance: why AM printing is used in transportation

As in the medical field, there are many different types of applications in the different areas of transport. However, one of the common characteristics of transport is that it involves the movement of an object and the energy (and thus cost) of moving an object is related to its weight. Weight reduction is an important benefit of AM (Gasman, 2019), which leads to *cost efficiency* of the use of the product. Reducing the mass of a Boeing 787 by 20% would improve fuel efficiency by 10%–12% while improving other performance characteristics (Blakey-Milner et al., 2021). Virgin Atlantic estimates that shaving a single pound off all the planes in their fleet would save them 14,000 gallons of fuel per year.[70] Weight reduction is perhaps the most obvious in aviation and aerospace, the latter of which is constantly demanding

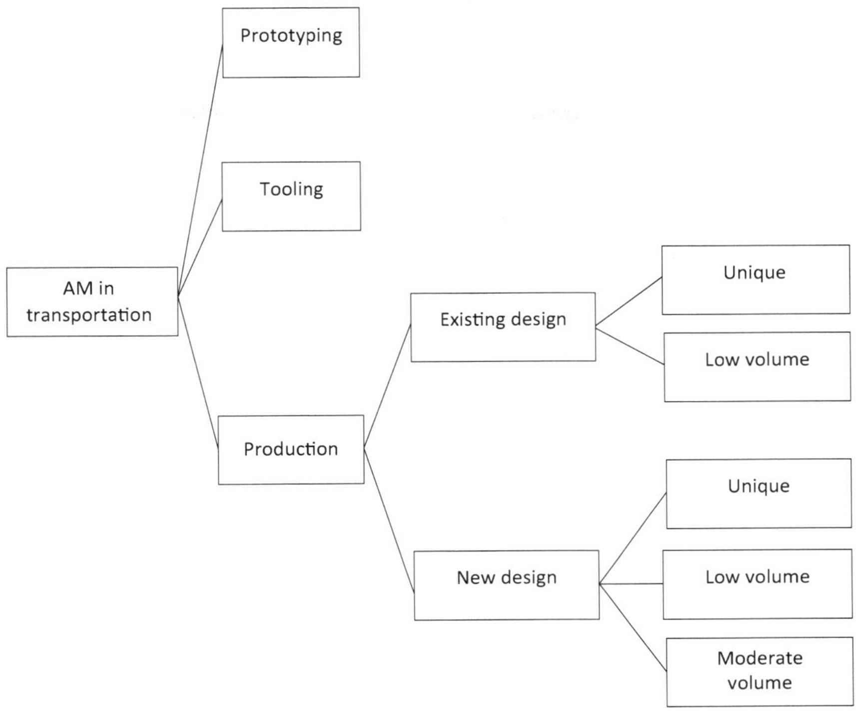

Figure 10.3 Use of AM in transportation.

Source: author.

lightweight components with a high strength-to-weight ratio (Najmon et al., 2019), but also applies to other types of transport. As discussed in Chapter 7, sending (construction) material into space is very expensive. An important reason to apply AM is that it can help to reduce weight by (a) simplifying/redesigning parts (less material), (b) through the use of lighter materials (which often are suitable for use with AM but less so for TM), and (c) by producing on site so that the cost of transport is limited. This latter aspect applies in particular to space exploration, so that items aren't transported but instead are manufactured in space and ideally with material available in space, such as on the moon or other planets. In terms of lighter materials, there are continual developments of new types of alloys and composites to accomplish this (see Froes and Boyer, 2019).

Weight and cost reduction in transport are connected to other things, as discussed above. Generally, in the context of one-of-a-kind or limited production runs, weight and cost reductions are achieved through (re)designing and customizing products (*product flexibility*) using functionally new and more complex shapes leading to part consolidation (*product innovation*), and by using AM, there is a reduction of the need for creating tooling, leading to a faster *time to market* and *delivery time*, and AM allows the use of new materials, leading to higher-quality parts (*product quality*) while lowering waste (*process sustainability*) (Najmon et al., 2019). An example is the GE9X heat exchanger, which had 40% mass reduction, 25% cost reduction and consolidation of 163 components into a single part when it moved from TM to AM (Blakey-Milner et al., 2021). A concept used in aviation is the buy-to-fly ratio, which is the weight ratio between the raw material and the weight of the final component. Some components have ratios up to 40:1, illustrating high waste, whereas with AM the ratio is much closer to 1:1 (Najmon et al., 2019). The same principle applies to part design for cars but doesn't have the same impact as in aviation/aerospace; this is because of the much higher production volumes in car manufacturing. Instead, prototyping remains the primary application of AM and AM is also widely used for tooling; see, for example, an interview with an automotive supplier.[71] Nevertheless, end-part production is also taking place for car manufacturing, although this is more likely to be with some type of plastic instead of metal. Figure 10.4 illustrates the use of AM for car prototyping. This was demonstrated at the 2022 FormNext tradeshow by Lincsolution. They work with car manufacturers to

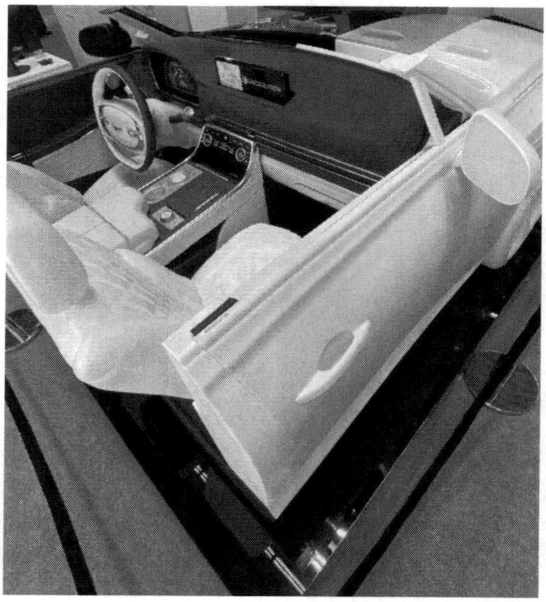

Figure 10.4 Part of a car prototype presented by Lincsolution at FormNext.

Source: author.

create prototypes. Traditional prototypes are made with clay, but switching to AM reduces the cost from $180,000 to $42,000 and reduces the time from 3 months to 2 months.

Although weight reduction is important, lead time reduction is currently the main benefit of AM for aerospace (Blakey-Milner et al., 2021) as production lead time has become a key indicator for competitiveness (Gisario et al., 2019). AM helps to reduce lead time in several important ways such as (a) use of material in powder form (which is readily available) compared with many aerospace alloys that have long lead times, (b) reduction or removal of specialty tooling and fixtures (tooling lines often take months), and (c) reduction in processing time, which is probably most important for spares or replacements that are out of production (Blakey-Milner et al., 2021).

Aside from this general pattern related to weight, *cost efficiency* is improved for relatively unique parts, such as for unique vehicles, because AM can be used economically for small volumes, thus limiting the need to produce large batches and keeping parts in inventory.

There are also new types of services (*service innovation*) such as automotive companies providing files to customers so that customers can print their own parts.

Although generally it does not appear to be a major concern, sustainable products are made in some instances; examples of this are boats that can get recycled at the end of life to create new boats (*product sustainability*).[72]

Another benefit of using AM in transportation is that it can involve expensive goods such as cars or airplanes, and speed may play an important role, as indicated above (*time to market* and *delivery time*). In many instances, tooling can be avoided when using AM, so that significant time savings occur because a mold doesn't have to be created. For example, owing to a late design change, GM needed a small part that was holding up delivery of 30,000 SUVs, so the company approached GKN Additive. If IM was used, then the tooling for IM was going to take 12 weeks. Because GKN Additive used an AM approach and AM doesn't need tooling, it was possible to deliver the parts in six weeks.[73] Another example of how AM can help the speed dimension is that of long-term aircraft storage due to grounding. This requires engine covers. During Covid, an airline like SAS didn't have inventory of off-the-shelf covers, and owing to Covid supply chain issues, ordering wasn't feasible. SAS used a remedial fix that was time-consuming (an 8-hour process of engine unpacking and repacking for weekly checks). A more permanent solution which required casting was designed, and the entire production process was able to go faster through 3DP of the tooling for this casting process.[74] Similar to Table 3.3, Table 10.1 summarizes the above discussion in terms of how AM affects the strategic dimension and related manufacturing capabilities for the transportation sector.

Table 10.1 Why AM is beneficial in the transportation sector

Strategic dimension	*Manufacturing capability*	*Transport*
Cost	Cost efficiency	Especially for customized low-volume items such as for luxury cars, race cars and rockets
	Resource efficiency	Not applicable. While productivity improvements are made, for example, by the use of tools that can easily be created with AM, this is already covered under product innovation.
Quality	Product quality	Products with better performance due to, for example, the use of different materials
Time	Delivery time	For small-volume production and also for situations when producing on location
	Time to market	Use of AM for prototyping and tooling
Flexibility	Product flexibility	Transportation is full of customization examples such as unique parts needed for old cars or rockets.
Innovation	Product innovation	Especially enabling new shapes in aerospace
	Service innovation	New services such as providing files to customers to print
Sustainability	Product sustainability	Less emphasis but examples exist of products that are made with specific materials for recycling purposes
	Process sustainability	Lower waste for complex products such as engines. A special case is 3DP in space and using available space materials.

Source: author.

AM methods that are used for transportation applications

Owing to the broad definition of transport used in this chapter, there is also a broad range of AM methods that are used. The types of AM that are used depend upon key characteristics such as whether a part is metal and the quality requirements. End-use parts have stringent strength requirements, whereas for prototyping, these requirements are less, but instead having color may be useful.

There are many instances where metal (end-use) parts are used in transport, as was discussed above. The main two types of AM used for metal parts are PBF and DED. PBF is generally used for small parts (sizes up to 400 mm), while larger objects are made using DED (Blakey-Milner et al., 2021). Transportation also often has plastic parts for end use. These are frequently made with PBF as well if there are strength requirements (Gasman, 2019). For items that are less sensitive to strength or that may not be for end use, other types of AM such as MEX and VPP may be considered. Another important application of AM is tooling such as VW's small jig for placing a logo. The quality requirement for this type of jig is less than for end-use parts and

Table 10.2 AM technologies used for transportation

	AM process						
	MEX	*PBF*	*VPP*	*MJT*	*BJT*	*DED*	*SHL*
transport	Common (e.g., tooling)	Most commonly used (especially metal in aerospace)	Prototype	Prototype	proto-type	Most commonly used especially aerospace	Uncommon

Source: author.

therefore MEX is a good option. The quality requirements for prototypes are perhaps the least restrictive in terms of strength of a part. They serve the purpose of showing something. Hence, types of AM such as MJT and BJT are also applied in transportation for prototyping purposes (Gasman, 2019). Table 10.2 provides a summary of the above discussion.

Where in the transportation value chain

In order to understand where AM is applied in transportation, it is necessary to understand the context. Transportation is considered a producer-driven chain (Gereffi, 1999). In producer-driven value chains, the manufacturer (e.g. companies such as Airbus, Boeing, Ford and Volkswagen) plays the central role in coordinating global production networks. The manufacturer is largely in control of the design of a product and then allocates production of parts to different parties that may be internal or external to the company (outsourcing). Companies in this production network fulfill different roles depending upon their capabilities and strategic reason for being in the production network (Ferdows, 1997). And, depending on the capabilities of outside suppliers, they may have more or less freedom, or a role, in parts design (Liker, Kamath and Wasti, 1998). Supplies typically end up with the manufacturer, which integrates them together into the final product (i.e., a system integrator) (Paasi et al., 2010). Abele et al. (2008) looked at the need for concentrating production in terms of cost and the need to disperse production in terms of market proximity and found five types of production networks: world factory, local for local, sequential or convergent, hub and spoke, and web structure. Within the same industry, different types of global production networks can coexist. Figure 10.5 provides a general view of a supply chain in the context of transport.

A. On the R&D side, system integrators use AM for prototyping or designing and making of production tools. This also applies to suppliers with expert knowledge. Examples are Volkswagen (automotive integrator) and Bosch (automotive supplier).

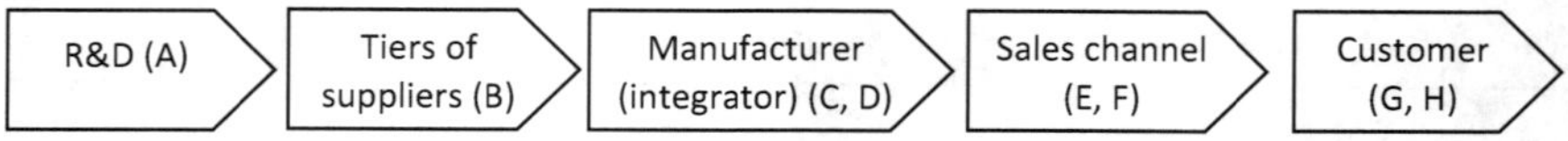

Figure 10.5 General depiction of a supply chain related to transport.
Source: author.

B. Different tiers of suppliers use AM for the production of parts that either become part of a larger component or are end-use parts. Examples of suppliers are BorgWarner and Visteon for automotive. This includes parts for specialty and low-volume customized products such as race cars and rockets. For example, SOLIZE is 3DP spare parts for classic Nissan models.
C. Similarly, the integrators and manufacturers use AM for the production of (end-use) parts. An interesting concept is the "mobile factory" such as used by Local Motors and Omniplus. This is essentially a production facility in a container that incorporates 3D printers so that it can be moved to a different location and produce close to demand.

The three types of uses above apply across different types of transport products and don't really change the business model. What they do is replace TM with AM. This change influences productivity because it can create parts with improved quality and/or for lower cost. Other than that, it doesn't dramatically change things.

D. Where things start to change is when AM is applied differently from the start, such as the design and 3DP of a customized bicycle such as by Arevo or an entire car such as the Strati. However, cars have existed since the late 1800s, and the industry is well established and has significant entry barriers. Success of a new entrant is by no means certain, hardly any start-up has survived, and it may take significant efforts and investments, as demonstrated by Tesla (Thomas and Maine, 2019).
E. Things also start to change at the level of the sales channel. This is because AM allows production to take place in a location closer to the customer (i.e., distributed manufacturing) (Srai et al., 2016). For automotive, this can be 3D printing of parts at the dealer level, although this has not yet taken off. The equivalent in aviation is maintenance, repair, and overhaul (MRO) facilities such as GE Aviation's Engine Services Singapore, which uses metal AM for commercial jet engine component repairs. In addition, specialized AM services such as Sculpteo, Shapeways, Treatstock and Materialise 3D print parts from existing designs, and this applies to other sectors as well. Some AM service companies help with design of new parts or redesign of existing parts.
F. Similar services exist where a digital design is sold so that the customer can 3D print it themselves. An example of this is iPrint3Dspares.
G. A variation on this theme is where the consumer becomes prosumer and 3D prints parts that may have been designed by the prosumer or found in other

ways such as through scanning or from a depository. There is an expanding set of a large variety of items varying from car interior accessories (e.g., a phone holder), exterior accessories (e.g., a hubcap) and functional other parts (e.g., a fuse box). Examples of websites that contain designs for car parts are 3Dmixers, GrabCad, MyMiniFactory and Thingiverse.

H. One step beyond this is where the prosumer 3D prints their own vehicle. This has already happened with a kayak, as discussed above.

Diffusion and challenges of AM in transportation

For aerospace, in the context of Figure 4.6, the concept of AM has been accepted, some specialized AM machines have been made and companies have adopted the machines for production purposes. To some degree, the AM concept has been adopted in the value chain, and products produced with AM have been accepted as well. It is also noticeable in transport that some companies have committed to AM through dedicated business units such as GE Additive and BMW's AM campus. In regard to the overall adoption in rough terms (specific products can be in slightly different stages), it seems that, in terms of Figure 4.7, AM in transportation is dealing with the chasm. This assessment is despite some surveys in aerospace indicating that up to 50% of companies are already using AM for the production of end-use parts (Metrix, 2021) or claims that AM is mainstream in aerospace (Gasman, 2019). Relevant here are how much of an aircraft is produced with AM, what type of parts are produced with AM, and what is the potential. For instance, Gasman (2019) indicates that production volumes in aerospace remain fairly low and that the primary application is prototyping and tooling. Safran illustrates this as it stated that currently less than 1% of its output features 3D printed components.[75] In general, some benefits have been clearly established, and companies have been adopting AM to get ahead in the market. However, Cotteleer (2014) identified four different pathways, but in aerospace most companies are in an early stage where AM is used for current products and without major changes to the supply chain structure (Coykendall et al., 2014). The challenge when crossing the chasm is to demonstrate that AM is well established and reliable and companies have a history of working with it. This is illustrated by a survey which found that executives in aerospace were looking for successful case studies before committing capital to AM setups (Coykendall et al., 2014: 14). This can also be seen in the communication from manufacturers about how long AM already exists, the cost benefits, their experience level and other things such as through cases on their websites. See examples for 3D Systems[76] (which has several cases and mentions de-risking AM application development) and Stratasys[77] (which includes a case study).

A key issue in transportation is certification such as for airplanes with the Federal Aviation Administration. These processes involve extensive testing for safety reasons and can be quite lengthy, which hampers the introduction of new things. For instance, during the 1950s and 1960s, fiber-reinforced plastics (composites) emerged as a new class of materials in aviation. Yet 30 years later, composite applications were limited to military aircraft, helicopters and some business jets. For civil aircraft, the main load-carrying parts were still made of aluminum alloys and in general around 80% of civil aircraft was still made from aluminum (Vlot, 2004). Thus, part of further diffusion is a matter of time. However, this is a continual process because 'new' AM processes (or with different characteristics) are frequently applied (in November 2021, PPG 3D printed parts for Lockheed Aircraft with a proprietary process called ambient reactive extrusion[78]) and new materials are continually developed (such as high-performance ceramics; Wang, Dommati and Hsieh, 2019). Thus, there is frequent reporting on testing of parts[79] that have been produced with AM technology and achieving certification such as CRP USA's certification for 3DP of parts for the aerospace industry[80] and GE Aviation receiving certification for its GE9X engine (with some 3D printed components).[81]

The viability of an AM business model in transportation, as in some other industries, can be uncertain. Local Motors is an example. They used AM for the production of parts for their Rally Fighter car that was introduced in 2009, 3D printed the Strati car in 2015, and in 2016 moved on to the Olli, an autonomous small bus. But, by January of 2022, Local Motors ceased operations. Part of the reason for this may have been that, in order to be successful for this type of vehicle, the company needed to sell a significant number of them. However, when the Olli became a more or less standardized vehicle, the advantages for using AM as the production technology may have disappeared. In contrast, a currently more viable path may exist for companies involved in space-related activities because this is a relatively new industry that is witnessing a change towards more private company involvement and lowering of cost because of the application of AM. The production quantities are smaller and frequently involve new, complex and unique designs that are a better match for where AM technology offers competitive advantages.

Notes

1 An impression can be found here: https://www.youtube.com/watch?v=aMQJkkchTi4
2 https://www.mlive.com/news/detroit/2015/01/worlds_first_3d-printed_car_fr.html
3 An informative video can be found here: https://www.youtube.com/watch?v=w22pNnyKl_M
4 https://www.tctmagazine.com/additive-manufacturing-3d-printing-news/local-motors-3D-printed-autonomous-vehicle-olli-deployed-berlin/
5 https://3dprintingindustry.com/news/olli-3d-printed-autonomous-electric-vehicle-transports-sacramento-state-students-149760/
6 https://www.digitalengineering247.com/article/urbee-could-be-the-first-3d-printed-car/

7 https://www.stuff.co.nz/motoring/84985936/a-3dprinted-electric-car-4ekolka-makes-its-informal-debut-on-a-prague-driveway
8 https://3dprintingindustry.com/news/chinas-1st-3d-printed-car-hits-the-road-45299/
9 https://www.motor1.com/news/101332/honda-micro-commuter/
10 https://www.tctmagazine.com/additive-manufacturing-3d-printing-news/divergent-microfactories-introduces-world-s-first-3d-printed-supercar-blade/
11 https://3dprint.com/293410/aston-martins-dbr22-features-fully-3d-printed-rear-assembly/
12 https://3dprint.com/270166/lamborghini-partners-carbon-rolls-out-sian-roadster-3d-printed-air-vents/
13 https://all3dp.com/porsche-3d-printing-metal-plastic-spare-parts-classic-cars/
14 https://www.fabbaloo.com/news/formlabs-focusing-on-automotive-applications
15 https://3dprint.com/288783/ams-speaker-spotlight-a-niche-auto-market-tailor-made-for-3d-printing/
16 https://3dprint.com/270300/mahle-trumpf-porsche-introduce-3d-printed-pistons-for-gt2-rs/
17 https://3dprint.com/270715/markforged-metal-3d-printing-replaces-obsolete-part-for-legacy-race-car/
18 https://3dprint.com/289955/3d-printed-car-accessories-give-new-peugeot-308-high-tech-flair/
19 Computer-Aided Design
20 https://3dprint.com/288885/2022-ford-maverick-cad-files-enable-3d-printed-accessories/
21 The companies need a certified 3D printer and a license for the required part in the required quantity. See https://www.3dnatives.com/en/daimler-buses-is-allowing-its-customers-to-3d-print-spare-parts-280520224/#!
22 https://3dprintingindustry.com/news/daimler-buses-creates-mobile-3d-printing-center-for-series-production-of-spare-parts-190011/
23 https://3dprint.com/292368/brumos-racing-finishes-1st-in-time-attack-thanks-to-airtech-3d-printing/
24 https://3dprint.com/291633/stratasys-announces-partnerships-with-nascar-toyota-to-3d-print-racecar-parts/
25 https://3dprintingindustry.com/news/interview-alpines-pat-warner-offers-a-unique-insight-into-the-secretive-world-of-3d-printing-in-formula-1-209139/
26 https://3dprintingindustry.com/news/eos-to-3d-print-race-car-parts-for-sauber-technologies-as-part-of-new-three-year-partnership-209734/
27 https://3dprintingindustry.com/news/to-the-moon-lamborghini-enters-metaverse-with-3d-printed-space-key-nft-202692/
28 https://3dprintingindustry.com/news/fresh-off-3d-printer-metal-parts-mercedes-benz-trucks-119701/
29 https://3dprintingindustry.com/news/toyota-materialise-team-to-3d-print-lightweight-car-seat-57779/
30 https://www.engadget.com/2017-08-07-michelin-vision-biodegradable-3d-print-airless-tire.html
31 https://ultimaker.com/learn/volkswagen-autoeuropa-maximizing-production-efficiency-with-3d-printed
32 https://www.fabbaloo.com/news/tata-group-and-3d-printing
33 https://3dprint.com/289555/volkswagens-ducati-to-use-roboze-3d-printing-on-superbikes/
34 https://3dprintingindustry.com/news/sturdy-cycles-leverages-headmades-cold-metal-fusion-technology-for-titanium-bike-parts-202634/
35 https://www.3dnatives.com/en/optimized-bicycle-saddles-amt-postpro-120920224/#!

36 https://3dprint.com/291826/first-load-bearing-metal-3d-printed-spare-part-for-airplanes-unveiled-by-lufthansa-and-premium-aerotec/
37 https://3dprint.com/269869/stratasys-aerospace-webinar-series-value-proposition-additive-manufacturing-to-airlines-presented-chuan-ching-tan-afs/
38 https://3dprint.com/256489/interview-with-scott-sevcik-vp-aerospace-stratasys-on-3d-printing-for-aviation-and-space/
39 https://3dprint.com/262742/boeing-777x-takes-first-flight-with-over-600-3d-printed-parts/
40 https://3dprint.com/273923/273923/
41 https://www.3dprintingmedia.network/ge-aviation-ships-100000th-3d-printed-nozzle-tip/
42 https://3d-prints.com/2020/01/28/boeing-77x-ge9x-engines-with-300-3d-printed-parts-powers-largest-twin-engine-jetliner-in-first-flight/
43 https://www.digitalengineering247.com/article/monash-university-3d-prints-complete-jet-engine/ and https://3dprintingindustry.com/news/orbex-raises-another-46-4m-ahead-of-first-prime-rocket-launch-in-the-uk-216468
44 https://3dprint.com/guides/?article_series=3d-printing-drone-swarms
45 https://3dprint.com/291640/nasa-funds-3d-printing-research-in-2022-sbir-sttr-awards/
46 https://3dprint.com/278887/3d-printing-and-the-future-of-space/
47 https://3dprintingindustry.com/news/fleet-space-develops-next-gen-3d-printed-satellites-to-bolster-iot-connectivity-200894/
48 https://www.cnn.com/2023/03/22/business/relativity-rocket-launch-florida-scn/index.html
49 https://3dprintingindustry.com/news/slm-solutions-3d-prints-landmark-single-piece-rocket-engine-for-orbex-148726/
50 https://3dprintingindustry.com/news/skyroras-3d-printed-rocket-engine-passes-latest-70kn-test-211108/
51 https://3dprint.com/58203/rocket-lab-electron-rutherford/ and https://3dprint.com/292360/rocket-labs-3d-printed-engine-powers-nasa-capstone-satellite-mission/
52 https://3dprint.com/291546/us-space-force-awards-launcher-1-7m-contract-for-3d-printed-rocket-engines/ and https://www.ursamajor.com/blog/ursa-majors-advanced-manufacturing-lab-in-youngstown-ohio-delivers-its-first
53 https://3dprint.com/292121/rpm-3d-printed-the-seats-for-boeings-starliner-spacecraft/
54 https://www.fabbaloo.com/2015/05/nasas-3d-printed-zero-gravity-coffee-cup
55 https://spaceflight101.com/iss/made-in-space-1st-iss-3d-printer/
56 https://www.space.com/made-in-space-3d-print-ceramic-turbine-part
57 https://inhabitat.com/engineer-makes-worlds-first-3d-printed-kayak-for-just-500-with-a-homemade-3d-printer/
58 http://depts.washington.edu/open3dp/2012/07/woof-rocks-the-boat/
59 https://bigrep.com/posts/amazea-underwater-scooter-pr/
60 https://www.3dnatives.com/en/large-format-additive-manufacturing-boat-building-repair-210920224/#!
61 https://3dprintingindustry.com/news/worlds-first-3d-printed-fiberglass-boat-to-debut-at-genoa-boat-show-2020-176547/
62 https://3dprintingindustry.com/news/umaine-develops-worlds-largest-3d-printed-boat-and-polymer-3d-printer-163224/
63 https://3dprint.com/288658/autonomous-underwater-vehicles-to-receive-3d-printed-parts-via-3dprintuk/
64 https://3dprint.com/278319/naval-group-uses-waam-to-3d-print-first-propeller-installed-on-french-navy-ship/

65 https://3dprintingindustry.com/news/u-s-navy-equips-uss-essex-warship-with-3d-printing-technology-212073/
66 https://3dprint.com/273462/breaking-ice-3d-printed-parts-refurbish-icebreaker-ships-for-arctic-sea-transport/
67 https://triblive.com/business/wabtec-corp-opens-new-3d-printing-operation-near-pittsburgh-airport/
68 https://www.3dnatives.com/en/applications-for-3d-printing-at-the-heart-of-the-railway-industry-110320224/
69 https://3dprint.com/292353/french-railway-leader-3d-prints-spare-parts-with-metal-filament-and-basfs-replique/
70 https://www.triplepundit.com/story/2014/airline-industry-shaves-weight-save-money/42651
71 https://www.3dnatives.com/en/bosch-interview200720184/
72 https://3dprintingindustry.com/news/tanaruz-plans-to-produce-300-3d-printed-boats-by-2023-207466/
73 https://3dprint.com/291836/3d-printing-for-gm-suv-opens-doors-for-gkn-additives-flexible-manufacturing/
74 https://bigrep.com/posts/jet-engine-covers-3d-printed-molds/
75 https://3dprint.com/294884/aerospace-giant-safran-opens-aerospace-e803d-printing-campus/
76 https://www.3dsystems.com/aerospace-defense
77 https://www.stratasys.com/en/industries-and-applications/3d-printing-industries/aerospace/
78 https://3dprint.com/287046/ppg-3d-prints-lockheed-aircraft-parts-with-new-proprietary-tech/
79 For example, NASA testing whether parts produced with AM can withstand the extreme heat that parts with traditional manufacturing can withstand; see https://3dprintingindustry.com/news/nasa-completes-23-hot-fire-tests-on-critical-3d-printed-engine-components-180912/
80 https://3dprint.com/273633/crp-usa-achieves-as9100d-certification-for-3d-printing-aerospace-parts/
81 https://www.fabbaloo.com/2020/10/ge-aviation-receives-certification-for-partly-3d-printed-jet-engine

Part III

Developments

11 Technological developments

Technological developments in AM happen frequently and are very diverse. One of the reasons for this relates to the diffusion of the technology (see Chapter 4). Broadly speaking, as shown in Chapters 6 to 10, much of the application of AM is in rapid prototyping and customized and small volumes (including tooling). One area that the industry is very much interested in is the ability to scale things, which means going to larger volumes of products such as successfully done with the GE fuel nozzle. In May 2022, the Joe Biden administration joined several large American companies to launch AM Forward. This is a public–private partnership with the purpose of increasing the use of AM by US-based small and medium enterprises. The increased adoption of AM remains challenging because of characteristics of AM processes, and, logically, there are many developments taking place to improve these characteristics. Some of these are broad in nature and involve core technology characteristics or limitations of the seven main AM types. For example, in VPP, acrylate-based plastics are used because these offer rapid curing and good spatial resolution; however, they also have low toughness and tend to be brittle. Newer machines are capable of dealing with thicker liquids which move away from acrylate-based materials and allow for stronger parts (Voet et al., 2018). There is even an expansion of 3D printing into four dimensions (Momeni, Liu and Ni, 2017).

Some developments are narrow in focus and improve a small part of AM capabilities such as increasing the build volume, dealing with leveling issues on ships, and making improvements related to extreme situations such as micro AM or micro-gravity AM. An example is the recent upgrade of the Ultimaker S5 to the S7. The latter provides faster AM with the capability to handle more and different materials.

Many developments fall in between the broad and narrow extremes, and it is impossible to keep track of all of these developments. The purpose of this chapter is to provide some general insight into how technological developments relate to the business case for AM. The AM technology consists of hardware, software, and the inputs to the AM process such as the digital files and materials. In terms of the improvements, these relate to the performance objectives and connected manufacturing capabilities. This theoretically leads

DOI: 10.4324/9781003399100-14

Table 11.1 Framework for AM technology developments

	Hardware	*Software and digital files*	*Materials*
Cost			
Quality			
Time			
Flexibility			
Innovation			
Sustainability			

Source: Author.

to Table 11.1, but it should be kept in mind that the types of AM have different advantages and disadvantages, so some cells may include developments that relate to one AM type but not be relevant for another AM type. There is also overlap between hardware, software and digital files and materials.

For the performance objectives and related manufacturing capabilities, the discussion will focus on the two most important capabilities: quality and speed. While cost is important and AM has to be feasible, AM is generally in an early stage with growth, which typically means that there is a widening buyer group with products that have technical and performance differentiation. Cost (and price) is more an issue in mature and declining markets (Porter, 1980). Improvements in the dimensions of quality and time affect flexibility and innovation, so these are not covered, and sustainability is discussed separately in Chapter 13.

Hardware

Developments in hardware concern the components of the 3D printer or combinations of components, including new AM machines. Many of these hardware developments require simultaneous software adjustments. Generally, the developments improve or extend the capabilities of AM and involve changes and nuances to the types of AM. Some of these developments are very broad and are at an early stage, such as an AM process that is based on sound (direct sound printing or DSP) which has the potential to 3D print directly inside a human body (Habibi et al., 2022). Another example is HIAM (hydrogel infusion additive manufacturing), an AM metal process that involves the infusion of hydrogel, which allows a resolution that is a magnitude smaller than previously possible (Saccone et al., 2022). Another development is volumetric 3DP, which is done by companies such as Vitro3D. This is a technique that makes it possible to solidify an entire object at once instead of layer by layer, which potentially makes the 3DP process 100 times faster. Cold spray technology, such as by SPEE3D, is another development. This is a fast metal AM technology that can use relatively low-quality and low-cost powder.

A little more nuanced are technology developments such as by Nexa3D, which is using lubricant sublayer photo-curing (LSPc). This is broadly similar to DLP but involves a new type of membrane (the part between the liquid and the light source) which has a no-stick zone. With DLP, once a layer is cured, the object is lifted up to allow for a new layer of liquid resin and then the object is lowered again. The process of lifting the object up and down is time-consuming, but Nexa3D's technology with a no-stick zone means that it can be done much faster. CLIP is another VPP process, and Stanford researchers have come up with a slight modification to the process that potentially can be five to ten times faster (Lipkowitz et al., 2022). Another interesting technology development is that related to the print head for MEX. Liqtra is a company that has developed a printhead that has seven nozzles (holes) (see Figure 11.1). This means that the print speed can be three to four times faster than for traditional MEX.

Still in development is work at Rutgers University on what is called multiplexed fused filament fabrication (MF3). This approach has multiple nozzles and allows (a) simultaneous movement such as producing parts in parallel or (b) working on different parts of the same object. The industrial F2 machine by Duplex 3D also uses two printheads, based on delta motion, and what is

Figure 11.1 Liqtra's seven-nozzle technology.

Source: author.

unique is that one is upside down so that one printhead can work on the top of an object while the other works on the bottom. Sculpman developed a nozzle with variable width. The advantage here is that certain parts of the print can be printed fast (but with less detail) while other parts can be printed with more detail (and slower). Another nozzle-related development is that of an induction hot end by Plasmics. Induction heating is potentially faster and provides more even heating than with traditional hot ends. At the Zurich University of Applied Sciences, they are working on a printhead that rotates. Traditional MEX has a weakness with overhangs and requires support structures but that can be overcome with this type of technology (Wüthrich et al., 2021). Other smaller improvements are build platforms made out of different materials to increase adhesion (e.g., Prusa's textured PEI powder-coated steel sheet). Another type of technology development is that by Xjet, which uses NPJ to make metal parts on their Carmel 1400 machine. With traditional MJT processes, there is human involvement in removing excess material, which requires handling chemicals. Xjet's technology does this inside the machine so that people don't handle the chemicals, which helps with safety improvements. AddUp provides a similar example of closed-off processes for PBF where operators do not handle (toxic) metal powders. Yet another type of development is that of specialized AM machines. For instance, Stratasys has developed specialized machines for the textile industry.

Software and digital files

One software development is for the underlying format that is used in AM, namely the STL file format. STL is based on triangles, which means that a 3D object is represented by triangular meshes. However, this also means that it does not fully represent an object accurately. Additionally, there is a trade-off between model quality and file size. An alternative is direct CAD slicing (Nayyeri et al., 2022). Although many companies still depend upon slicing software that is based on STL, a company such as GrabCAD provides slicing software that uses direct CAD slicing. Another interesting development in slicing software is that of Markforged. They added a feature to their Eiger slicing software, which uses simulation to adjust the slicing. For instance, when planning to 3D print a part, the simulation can point out that certain areas need to be stronger than other areas, and with this information, the slicing process can be adjusted and improved. A big issue with the software consists of developments that deal with AM production from start to finish, including design editing and post-processing. All of this needs to function together and potentially across different factories. Oqton is one of the companies working on this, by applying artificial intelligence.

There are also developments in the area of the creating models through the development of more advanced or less expensive scanners. The Einstar by Shining3D was introduced in 2022 and is a relatively affordable (less than $1,000) and highly functional scanner that comes with model processing software. A related development is that of scanning capabilities through mobile phone apps such as Trnio. There are also attempts to use artificial intelligence for the generation of 3D printable models, but these are in an early stage of development.

Another development related to 3DP is that of platforms where files can be downloaded and thus provide value in terms of availability of models. Some potential legal issues with these platforms will be covered in Chapter 13. Additional developments include the Digital Design Warehouse by Authentise, whose goal is to offer a catalog of spare parts.

Materials

A core area of AM developments is materials. This is to be expected because materials play a huge role in what can be made with AM and the technical specifications. News about materials falls into three categories. First, there are completely new materials that were invented specifically for 3DP. An example of this is GRX-810, a new metal 3DP alloy that was specially designed by NASA for use in high-performance aerospace systems and that is up to 1,000 times more durable than existing aerospace alloys. Another example comes from Carbon, which developed a new bioresorbable material that could be ideal for applications in tissue repair.[1] Yet another example comes from Lithoz, which developed Silicon-Infiltrated Silicon Carbide (SiSiC), a lightweight but hard ceramic material. Universities are also involved: Amherst University and Georgia Institute of Technology together developed an ultra-strong and ductile metal alloy (Ren et al., 2022), and a team of researchers from Monash University developed an ultra-strong titanium alloy (Zhu et al., 2022).

Second, some previously known materials have become 3D printable. An example is silicone, which is used in IM, but owing to its characteristics, it is difficult to 3D print. Nevertheless, companies such as Spectroplast, a spinoff of the Swiss Federal Institute of Technology (ETH) in Zurich, has developed a machine that can produce silicone parts with dimensions that are not possible with IM. ETH Zurich also does research on using salt for 3DP molds (Kleger et al., 2022). Other examples come from Carnegie Mellon (Garg et al., 2022) and the Indian Institute of Technology Bombay (Kamble et al., 2022). Both teams developed a method to 3D print ice. Sometimes, the materials are slightly altered to make them feasible for 3DP. An example is the use of carbon, which strengthens the product. Markforged is a company that combines carbon fibers in their materials so that very strong parts can be produced with MEX. Another example is 3DP of glass, which is difficult to process. Syngis has developed a machine that, in combination with a new type of low-temperature glass, can 3D print glass objects. Yet another example is that of jute, which can now be 3D printed (Devajaran et al., 2022).

Third, there are developments with regard to other alterations that are less technical in character. For example, the number of colors of PLA material, which is often used with MEX is growing. Companies such as Rosa 3D, Solvay and Spectrum Filaments produce a large variety of materials and colors for 3DP purposes and keep coming out with new materials. At FormNext 2022, Spectrum Filaments presented eight new technical materials such as composites of kevlar, teflon, carbon fibers, and glass fibers.

Product quality

The quality of 3D printed products is important for the diffusion of AM. Size is among the dimensions that receives attention, as is color. For example, while BJT easily allows color but MEX does not. XYZprinting's approach with its da Vinci to this has been to infuse color into a color-absorbing filament, while Prusa's approach has been to have five filaments connected to one extruder and swapping between them during the print. A disadvantage of this is that when the color is switched, the old color is purged from the extruder, leading to waste. An alternative approach is to use multiple nozzles and/or multiple extruders, which is what Flashforge's Creater Pro 2 and INTAMSYS FUNMAT Pro 310 do (multiple extruders, in the latter case independent), and Multec's M800 uses six hot ends.

Tracking quality is another important issue where developments take place. An example of this is a scanhead in PBF that allows tracking of quality during the production process.[2] Some manufacturers have started to provide quality specifications with their AM-produced parts.[3]

Below are a few examples of quality improvements that offset weaknesses in a particular AM process:

- At the University of Amsterdam, a hybrid process of DLP and SLA which offers high resolution and high speed was developed.[4]
- Two-photon polymerization (2PP) technology by Vienna-based UpNano is an AM approach where a laser very precisely targets a specific voxel of resin. This gives it the advantage of being able to print extremely detailed elements. However, it can also enlarge the focal point to harden areas quickly, thereby combining the ability to print detail with that of providing a speed advantage.
- PBF has disadvantages such as high cost and low throughput, while a disadvantage of BJT is the traditionally weak parts. GE's metal BJT series 3 machine uses a different BJT approach that allows for the production of strong parts. This makes GE's BJT competitive with PBF.
- Visitech's Direct Image Sintering (DIS) approach is similar to the idea of DLP where an entire layer is cured at once, except Visitech uses this approach in PBF where it is able to produce stronger parts, faster and with larger build volumes than traditional SLS.
- Aerosint's Selective Powder Deposition (SPD) is a PBF-oriented approach that allows multi-metal 3DP.

Delivery time

While delivery time is influenced by more than just processing speed (e.g., it is influenced when using print farms; see Chapter 3), the speed of AM processes is an area where major developments occur. Faster AM processes are appealing because customer orders can be completed in less time and more work can be done on AM machines per time unit. This lowers the machine cost per part (see Chapter 3). Unfortunately, the speed of AM processes isn't as clear-cut as one might think. For example, a desktop MEX machine specification might provide an indication of print speed; in reality, this isn't a fixed number. First, the material that is being printed influences this speed. TPU can't be printed as fast as PLA. Second, the time to complete a print depends upon variables such as nozzle size, layer height, infill percentage, and whether supports are used. Third, the speed can vary depending upon the task such as the first layer, filling in the object, bridging, and the perimeter. Nevertheless, speed receives a lot of attention, and there are several developments that help to increase the overall speed of AM. VPP examples such as Nexa3D's LSPc process were already discussed above. An additional promising approach that is under development and has the potential to increase the VPP speed is that of using two lasers, one that is blue and one that is red(Hahn et al., 2022).

For PBF, the number of lasers is considered as a way to improve speed. Nexa3D provides a PBF machine which has four lasers to increase the speed. In 2022, Farsoon followed this example and announced its four-laser HT1001P machine. A metal PBF machine such as the Sapphire XC from Velo3D has even more lasers: eight. In 2022, EPLUS 3D introduced its EP-M1250, a large metal 3D printer with nine! For MEX, a similar approach can be seen with dual extrusion machines. Another technologically potential approach with PBF (pursued by, for instance, NUBURU) is to use a blue laser instead of an infrared laser. Blue lasers can provide high part densities, higher printing speeds, and a means for scaling up production speeds with multiple parallel lasers.[5]

For MEX systems, some of the barriers to higher speed are the ability to push material through the machine faster as well as have faster printer movements. Veda 3D has come up with VXS (Veda extrusion system), a new way of extruding which can be added to existing MEX machines and which is significantly faster than traditional extruders. Another intriguing solution, CIM (cast in motion), comes from Massivit 3D, which is a MEX process that uses light for curing. In this process, one head, using a water-soluble material that quickly hardens under ultraviolet light, extrudes an outer shell at high speed. This shell essentially functions as a mold. A second head extrudes a thermoset material inside this shell as it is being built. When the print is completed, it is cured in an oven to strengthen the inside thermoset material and after that the shell is dissolved in water, which leaves a quickly produced part.

Notes

1 https://3dprintingindustry.com/news/carbon-bioresorbable-elastomer-shows-biomedical-3d-printing-potential-216421/
2 https://3dprint.com/295265/new-scan-head-enables-in-process-quality-control-for-l-pbf-metal-3d-printers/
3 https://www.fabbaloo.com/2020/03/checking-out-minifactorys-3d-print-technical-sheets
4 https://3dprintingindustry.com/news/atum3d-licenses-hybrid-dlp-sla-3d-printing-technology-from-the-university-of-amsterdam-215512/
5 https://www.laserfocusworld.com/laser-processing/article/14279713/blue-industrial-laser-brings-speed-quality-and-efficiency-to-metal-3d-printing

12 Sustainability developments

With issues such as global warming, there has been increasing awareness that planet earth seems to be out of balance. At the company level, this has led to increasing attention on the role of companies in sustainability (Gupta, 1995; Longoni and Cagliano, 2015), and at the country level, international agreements, such as the Paris climate agreement in 2015, have been reached. One of the objectives of this agreement is to hold average global warming to well below 2°C above pre-industrial levels and to pursue efforts to limit the temperature increase to 1.5°C. This is important because rising temperatures affect a number of things, including animals (e.g., fish [Alfonso, Gesto and Sadoul, 2021] and insects [Warren et al., 2018]), growing crops (such as grapes [Venios et al., 2020] and rice [Chen et al., 2020a]), consequences of floods (Tabari, 2020) but also drought (Vicente-Serrano et al., 2020), and more generally the impact on ecosystems (Prăvălie, 2018).

In this context, AM offers a number of advantages compared with TM that can help to make progress on sustainability (Nadagouda, Ginn and Rastogi, 2020). AM influences how products are designed, and AM can help create designs that use fewer resources. Because of how AM takes place, it also has the potential to create less waste. Plus, on the material side, there are new materials that allow for more productive use of resources and improve recycling.

Despite the widespread use of words such as environmental sustainability, it is actually poorly defined, and there is a lot of scientific discussion about what it actually means (Morelli, 2011; Sutton, 2004; Virtanen, Siragusa and Guttorm, 2020), how to approach it (Goosen, 2012) and how to measure it (Kwatra, Kumar and Sharma, 2020). In a broad sense, environmental sustainability is viewed as having three components: an environmental/ecological part, an economic part, and a societal/human part (Duran et al., 2015; Marchese et al., 2018). AM can impact all three; for instance, on the society level, new products that benefit older people or people in poorer locations in the world can be developed. An example is new medical devices that improve the quality and length of life (which also connects with the economic side of environmental sustainability) or how 3DP of houses can help with affordable homes. This also gets into affordability and utility bills. Research on 3DP for

DOI: 10.4324/9781003399100-15

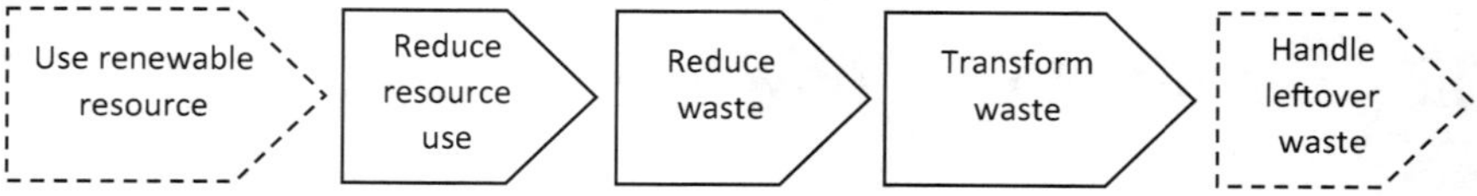

Figure 12.1 Environmental sustainability flow.

Source: author.

zero energy homes is ongoing (Sanguinetti et al., 2019), and Building for Humanity is using 3DP homes to tackle the problem of homelessness.

The main impact of AM is on the environmental/ecological dimension. Glavič and Lukman (2007) provide an interesting way of looking at the environment and ecological dimension which more or less follows the supply chain as depicted in Figure 12.1 and which is followed in the next sections.

Renewable resources

This first aspect from Figure 12.1 has to do with using renewable resources rather than resources that become depleted, such as fossil fuel. Although the carbon footprint is something that is affected by fossil fuels, it is not dependent upon fossil fuels only; therefore, this topic is covered in the natural ecosystems section towards the end of the chapter. The use of renewable resources is less relevant for AM because AM machines still require energy and it depends on the situation whether renewable energy resources are used. However, progress is made towards reducing the dependency on fossil fuels. For instance, Aalto University in Finland is developing new sustainable, bio-based plastic materials to replace conventional fossil-based plastics and composites.[1] A related concept is that of zero net energy buildings. Buildings consume about 30%–40% of the total energy consumption in the world and contribute to around 30% of the world's CO_2 emissions (Alkhalidi and Hatuqay, 2020). There are different viewpoints and definitions for zero net energy buildings (see Deng, Wang and Dai, 2014; Marszal et al., 2011; Wu and Skye, 2021), but the general concept of net energy is that of a balance between energy used versus generated by a building. More specifically, it relates to the energy balance of a grid-connected building, so that net-zero means that the balance between energy used and energy sold is zero (Hernandez and Kenny, 2010).[2] The company Mighty Buildings 3D printed a net-zero home.

Additionally, there are some examples on an individual basis of people who use renewable resources for AM, such as German designer Markus Kayser, who built a 3D printer that uses sunlight as the source of energy and sand as the printing material to make glass objects in the desert.[3]

Minimization of resource usage

If depletable resources are used, then this second aspect is about minimizing the use of such resources. This means the use of raw materials but also water and energy and other natural resources. Conservation of these resources improves their availability and maintenance; this is important for the survival of ecosystems. Furthermore, reduction in resource usage can lead to lower cost. Even though AM processes may use more energy, AM overall improves the minimization of resource usage (Khosravani and Reinicke, 2020; Shuaib et al., 2021). For example, AM processes essentially use just the materials that are needed. In some situations, such as selective laser sintering (SLS) processes, the production takes place with powder beds, but the powder that is not used in the AM process is used again for the production of another item. Companies such as start-up Simplyfiber use biodegradable materials with non-toxic additives so that waste is reduced. Food is an example where minimization of resource usage can apply. Meals can be customized so that they include those nutrients that are needed, reduce or eliminate undesirable items such as anti-nutritional factors, and contain the quantities appropriate to specific people. This reduces the use of unnecessary resources. In the construction industry, there is ongoing progress towards 3DP buildings that save energy (He et al., 2020) and require fewer resources, such as by using recycled sand instead of natural sand to make concrete (Ding et al., 2020). Owing to the development of new materials, it is possible to use lighter materials that are easier to process and cheaper than TM, such as what is done by Swiss company Schindler in the elevator industry.[4] Another area in which resource usage is reduced is that of tooling and prototypes. Producing prototypes or tools with AM compared with TM can reduce the amount of materials. An example is Pepsi, which changed to AM and found that the cost per tool set decreased from $10,000 to $350.[5]

Source reduction

The third aspect looks at source reduction, meaning reducing the quantity of materials that enter a waste stream. This includes dematerialization, which means quantitatively reducing the volume of material and energy used (while still meeting the user's demand and maintaining the same quality). This is an area where AM excels. In Chapter 6, it was pointed out how wasteful the fashion industry can be but AM can help to reduce waste. For instance, sunglasses have been produced that have no waste.[6] In Chapter 7, it was pointed out how wasteful the construction industry is because of its production methods, which include a lot of materials that are discarded after a building is complete. Using AM in construction makes a major contribution to resource reduction.[7]

Recycling, reuse and repair

If waste is created, then the fourth and the fifth aspects (the next section) look at the possibility to somehow use the waste. This fourth aspect looks at using the waste for production of the same or similar items as before, and the fifth aspect (regeneration, recovery and remanufacturing) is about using waste for production of other items. Using waste to produce the same or similar items

Figure 12.2 Machine to make filament from recycled plastic by 3devo.
Source: author.

can be done by recycling or reuse (where the waste becomes an input; with recycling it is accompanied by structural changes, while with reuse it is not) and repair (where the 'waste' is improved to increase quality and usefulness, essentially extending a product's life).

This is another area where AM excels. A key part of AM is the material that is used for the 3DP of products. Processes that involve MEX often involve filament that is some type of plastic. This filament can be created by recycling existing plastic. This can be done on a large scale but may also be beneficial on a smaller scale such as for prosumers. A company such as 3devo sells machines that create filament for 3DP out of plastic pellets (typically new plastic) or shredded plastic (recycled) (see Figure 12.2). It also sells machines for shredding so that people who buy both machines have the ability to, for example, shred orange juice bottle caps (typically colored orange) and then use the shredded plastic to create (orange colored) filament. It should be noted, however, that some uniformity in the plastic that is recycled is required, and it may require dehydrating for which 3devo also has a machine. Other companies that offer these types of machines are Filabot and Felfil. Polyformer is an open-source project for this type of machine.[8] What this also means is that companies or prosumers have the ability to recycle 3D prints that go wrong to reprint the part.[9]

Another example of this is the use of recycled uppers for 3D printed shoes by Adidas.[10] Sometimes, products or parts of products break, while replacement parts are not readily available, and these items end up getting discarded or not used. AM provides a method to help with this. Chapter 10 discussed how AM is used for making spare parts for cars, which helps prolong a product's life and reduce waste. A project between Bentax and the Danish Technological Institute provides another example and involves a key component of coffee machines that has a relatively high percentage of failure but, because of the TM cost involved, made little sense to be replaced. However, with AM, the costs are more reasonable, so the machines can be repaired.[11]

Regeneration, recovery and remanufacturing

This fifth aspect is similar to the previous section in the sense that waste is transformed so that it can be used. Regeneration is an activity of material renewal to return it in its primary form for usage in the same or a different process, and recovery is a process of restoring materials found in the waste stream to a beneficial use which may be for purposes different from the original use. Remanufacturing is the substantial rebuilding or refurbishment of machines, mechanical devices or other objects to bring them to a reusable or almost new state. Similar to the previous section, the machines by companies such as 3devo, Filabot and Felfil can be used to create filament that is used for 3DP a product different from what the material was initially used for. A practical demonstration of this is the recycling of plastic milk jugs to create filament that was subsequently used to 3D print a small boat.[12] Regeneration is also used by Pepsi, whose plastic bottles are used to create filament that can be used for 3DP.[13] Similarly, a partnership of US-based 3DP company Essentium and Brazil-based petrochemical company Braskem uses plastic from previous products (bottlecaps) for the production of filament for 3DP.[14] Roboze, active in 3DP, and Pielleitalia, active in textiles, are working on recovering polyamide clothing and accessories to transform them into technical filament for 3DP,[15] while Ford is regenerating material waste from 3D printed components into plastic pellets which are then used in injection-molding processes.[16] Research is taking place on upcycling, in which waste is blended with certain materials so that new filament is created but with more desirable characteristics (e.g., combining with carbon to make it stronger). Examples of this research are studies from the UAE University (Al-Mazrouei, Al-Marzouqi and Ahmed, 2022), Oak Ridge National Laboratory (Kim et al., 2022) and Washington State University (Shao et al., 2022).

Purification, end-of-pipe and degradation

The sixth and seventh aspects are the last two related to the environment and have to do with activities at the end of Figure 12.1 which are more about waste treatment and less relevant for AM.

To be complete, the sixth aspect consists of purification and end-of-pipe, which are about reducing the direct release of pollutants. Purification is the removal of unwanted mechanical particles, organic compounds and other impurities, while end-of-pipe is the practice of treating polluting substances at the end of the production process when all products and waste products have been made and the waste products are being released. The last aspect of the environment is degradation, which is a biological, chemical or physical process which results in the loss of productive potential.

Relationships between natural ecosystems

In addition to these environmental aspects of sustainability, there is an aspect that deals with ecology, which has to do with relationships between natural systems. For the purpose of the discussion, what is included here are examples of rejuvenating nature. For instance, bees substantially contribute to food security, medical resources and soil formation (Klein et al., 2018), such as through crop pollination (Heard, 1999). Unfortunately, owing to, among other things, pesticides (vanEngelsdorp et al., 2009), there has been a decline in bee colonies, which is known as colony collapse disorder (Rucker, Thurman and Burgett, 2019). AM has been used in Germany,[17] England[18] and Scotland[19] to 3D print beehives to help restore the bee population. Another example is cod that has suffered from overfishing (Brander, 2018), and owing to a lower cod population, the green crabs population has increased, causing a decline is certain seagrasses (Matheson et al., 2016), which provide natural shelter for young cod fish. AM has been used to help restore the cod population by 3DP artificial reefs that mimic natural shelters, giving the fish an ideal place to hide and nest and thus improving their chances of survival.[20] 3D printed artificial reefs are also used to foster coral reef development by providing spaces for coral to latch onto and grow, such as in the Maldives[21] and Hong Kong.[22]

AM is also used to help reduce the carbon footprint. The carbon footprint is defined as the quantity of greenhouse gasses expressed in terms of CO_2 emitted into the atmosphere by an individual, organization, process or product or even from within a specified boundary (Pandey, Agrawal and Pandey, 2011). As an aside, although it is clear that the rise in CO_2 concentration is due to human activities, what is less clear is the exact influence of humans on climate since there are many other variables involved (Koonin, 2021). In the footwear industry, a collaboration between Covestro, Huafeng and Cyclone uses special recycling processes and bio-based raw materials which significantly reduce the carbon footprint compared with traditional shoe production.[23] In the construction industry, construction for carbon-zero housing is taking place, such as in the UK.[24] Research is taking place on further developments, such as at Texas A&M University on carbon net negative construction through the use of hempcrete in 3DP buildings,[25] while at the University of Virginia[26] and the University of Tennessee and in Nanjing (China), research is taking place on 3DP walls and roofs where plants can grow (living walls), which helps to create carbon-neutral buildings (He et al., 2020).

AM and sustainability

What the above discussion shows is that the concept of environmental sustainability has multiple dimensions. In the context of AM, the most important dimension is that of the environment and connected to it ecology. There are many examples of how AM has helped to create a more sustainable environment. The two most prominent aspects of this are first the reduction of waste (a consequence of the additive nature of 3DP) and second, when waste is created, the possibility of transforming waste so that it can be used again (e.g., plastic that is turned into filament).

Notes

1 https://www.3dnatives.com/en/valuebiomat-3d-printing-biosourced-composites-120920225/#!
2 In some instances, the concept is considered in a broader context where the building is not connected to the grid, see (Marszal et al., 2011; Wu and Skye, 2021)
3 https://www.dezeen.com/2011/06/28/the-solar-sinter-by-markus-kayser/
4 https://www.fabbaloo.com/news/2022-3d-printing-and-the-global-elevator-industry
5 https://www.3dnatives.com/en/pepsi-relies-on-additive-manufacturing-for-bottles-070620224/
6 https://3dprint.com/291674/sustainable-3d-printed-luxury-sunglasses-collection-launches-with-zero-waste/
7 See for instance: https://3dprint.com/291613/indian-cements-and-tvasta-to-collaborate-on-3d-printed-housing/
8 https://www.reiten.design/polyformer/process
9 Whether this is economical is another issue because that will depend upon how much material is used and the cost of the machines.
10 https://3dprint.com/292717/latest-adidas-3d-printed-shoes-feature-recycled-uppers/
11 https://www.fabbaloo.com/news/broken-coffee-machines-saved-by-3d-printing
12 http://depts.washington.edu/open3dp/2012/07/woof-rocks-the-boat/
13 https://www.fabbaloo.com/news/pepsi-to-produce-3d-printer-filament
14 https://www.fabbaloo.com/news/essentium-announces-major-recyclable-materials-partnership
15 https://www.3dprintingmedia.network/new-collaboration-recycling-of-polyamide-for-3d-printing/
16 https://www.fabbaloo.com/news/ford-blue-oval-ev-plant-and-3d-printing
17 https://www.3dnatives.com/en/hiive-on-helping-bees-using-3d-printing-220220224/
18 https://3dprintingindustry.com/news/researchers-use-3d-printed-bee-nests-to-boost-uk-bumblebee-population-195362
19 https://www.scotsman.com/news/environment/scottish-conservationists-create-worlds-first-3d-printed-hive-to-help-save-honey-bees-3671299
20 https://3dprint.com/292843/3d-printed-reefs-aim-to-restore-cod-in-scandinavian-waters/
21 https://3dprint.com/294080/what-is-the-outlook-for-restoring-coral-reefs-through-3d-printing/
22 https://3dprint.com/292649/3d-printed-terracotta-stools-helping-to-conserve-the-coastal-shoreline-in-hong-kong/

23 https://www.3dnatives.com/en/can-3d-printing-reduce-the-carbon-footprint-of-footwear300520224/#!
24 https://www.3dnatives.com/en/building-for-humanity-affordable-3d-printed-homes-111020226/
25 https://3dprint.com/292638/arpa-e-funds-hempcrete-for-construction-3d-printing-via-texas-am-grant/
26 https://eandt.theiet.org/content/articles/2022/09/researchers-3d-print-walls-that-can-grow-plants/

13 Legal developments

As with any type of new technology, the laws may not be in place yet to deal with the newness and its consequences. This chapter describes some of the legal issues that are going on with AM. These are divided into four types of issues dealing with the product (what is produced and sold), ownership of AM processes, safety, and scams involving AM.

Legal issues with products

On the one hand, AM machines are like other large machines used in industrial production processes. On the other hand, some AM machines, particularly MEX, have a relatively small cost, which makes it a mini-factory for households. Thus, more so than since the first industrial revolution when manufacturing was moved into large-scale facilities, consumers now have the straightforward ability to produce goods. This also means that there is the possibility that something illegal is produced (Depoorter, 2014). This issue is exacerbated by the ease with which a digital file for AM can be obtained. In some cases, a designer can relatively easily recreate or reverse-engineer an existing product to create a digital file. An example is somebody who designed his own Crocs shoes with small modifications.[1] A digital file can also be obtained by using a scanner. Owing to 3D scanner technology, ordinary citizens, even those without design skills, have the capability to copy an object.

Property protection is a key factor in economic growth (Bernstein, 2004; Ferguson, 2011). It provides a mechanism for people to earn income for their ideas, and especially when it involves large R&D investments, it provides protection for companies so that they can get returns from this investment. A duplicate or copy of the design can avoid much of the R&D investments so that copied or reverse-engineered designs may be a lot cheaper. Therefore, without property protection, it may be difficult for companies to recover the R&D investment or for individuals to earn money for their ideas. However,

DOI: 10.4324/9781003399100-16

whether intellectual property (IP) protection works depends upon at least three factors related to enforcing it:

1. It depends upon the country and the strength of IP laws in that country whether pursuing IP violations is feasible.[2]
2. Even if IP law is strong, what matters is whether the IP owner has the financial means and time to litigate. The average cost of patent litigation in the US is $2 million, and the average time involved is 2 years (Jabaly, 2010). This may not be feasible for small entrepreneurial firms.
3. Even if the IP infraction can be defended, whether doing so is worth it depends upon the potential benefits, such as preserving reputation or receiving payment.

For instance, let's say that Disney sells plastic Mickey Mouse figurines for which it owns the IP rights and thus an ability to earn money based on the design of this cartoon character. With TM, these types of figurines can be produced with an IM machine. These machines are expensive and not feasible for a household to acquire. Companies that have such machines and that do not work with Disney are unlikely to use IM machines to illegally make Mickey Mouse figurines. If they were to do so, the company would need to make a mold for the figurine, which would take time and effort because it requires the design measurements of the Mickey Mouse figurine. Since the company would not work for or with Disney, they would have to reverse-engineer the Mickey Mouse design and then create a mold. To recover the high cost related to creating the mold, the company would have to produce and sell a large amount of the illegal figurines, and since it has to operate at scale, it may attract attention and Disney could sue the company. In this situation, the IP that Disney holds and the means Disney has to sue potential infractions discourage companies from breaking the IP law.

AM technology changes this scenario because it moves manufacturing to the realm of prosumers. A prosumer can purchase a relatively cheap MEX printer and scanner. It can use a sample Mickey Mouse figurine, scan this figurine and create a digital file. Thus, it becomes relatively easy for ordinary citizens to copy things and 3D print them. They could do this for a single item for personal use which would make it very difficult for Disney to catch this illegal activity (Brean, 2013). Even if a company like Disney would catch the illegal activity, it is unattractive to sue large amounts of prosumers for these infringements when the financial resources of prosumers to pay damages are limited (Bechtold, 2015). As a side note, the issue of legality goes both ways: Disney has been accused of illegally copying a design from Thingiverse.

Legal action can start in the form of a threat. During early Covid in 2020, when Italy was one of the countries hardest hit, there was a shortage of respirator valves in Italy. Issinova came up with a solution and 3D printed valves for a local hospital for less than $2 compared with the original $11,000 for the part. This led to a threat of a lawsuit, although this threat seems not to have been acted upon.[3]

When actions go beyond threats, key questions are: what is actually owned and by whom? During the 2015 halftime Superbowl show of Katy Perry, one of the background dancers in a shark costume was out of sync and received a lot of attention. This prompted somebody to design a shark figurine similar to the one on the show and make this available on Thingiverse. This led to a cease-and-desist order by Katy Perry's legal team because of infringement of IP. However, the USPTO rejected the claim that Katy Perry had design ownership.[4] Another example comes from a French museum that 3D scanned its sculptures. The French government concluded that these 3D scans were subject to public disclosure.[5] This gives prosumers access to the digital files and thus the ability to replicate these pieces of art.

Owing to the ease of sharing files on file repositories, these have also seen legal action (Bechtold, 2015). The first Thingiverse takedown notice was in 2010 (Rideout, 2011). Of particular concern are (spare) parts such as for a car (Ballardini, Ituarte and Pei, 2018), as demonstrated in 2022, when Honda issued a takedown notice of all Honda-related 3D printable files on Prusa's repository. Spare parts are challenging for a company to provide because demand is unpredictable, the parts need to support older generations of products that may have become obsolete, and stock-outs of critical parts disrupt operations, so that it requires companies to hold a relatively large inventory of parts close to where demand may occur (Holmström et al., 2010). AM has the potential to improve the spare parts situation for companies by offering the ability to produce closer to the end point and when needed (Akmal et al., 2022; Bhattacharyya, Kulkarni and Mishra, 2022). However, there are complications. First, not all spare parts can be replaced by AM parts. Studies have shown that less than 10% of spare parts can be replaced by AM; this is due to how parts were originally designed, the materials needed, manufacturing-related information that may be incomplete, and economic feasibility (Heinen and Hoberg, 2019; Knofius, van der Heijden and Zijm, 2016; Salmi et al., 2018). Second, whether distributed AM is the best option depends on, among other things, the acquisition cost of AM equipment and personnel costs versus the inventory and transportation costs (Khajavi, Partanen and Holmström, 2014). Third, in the context of patent law for commercially AM-produced spare parts, the making or "repair" via spare parts can violate a patent, but there is, at least in Europe, legal uncertainty related to what falls under legal repair or under illegal make (Ballardini, Ituarte and Pei, 2018). Fourth, in terms of tort liability, the legal ramifications are somewhat vague because traditional product liability is tied to the manufacturing function, but when parts are 3D printed by non-traditional sources, it is unclear who is liable when the product or part fails and there can be a variety of causes such as issues with the design, the process of sharing of the design, the material used in the 3DP process, or a defect in the 3D printer (Beck and Jacobson, 2017). This is even more complicated in the context of medical devices and health care. In 2022, a medical doctor was found criminally liable by a Swedish court because some of his patients became disfigured and nearly all of his 20 patients died after receiving his 3D printed artificial windpipes.[6] Work is ongoing about establishing better guidelines such as by the US Food and Drug Administration (FDA), which posted a discussion paper in 2021 for the purpose of generating feedback for future regulations.[7]

The previous examples have focused mostly on the company side of things, but the same mechanisms apply to individuals. An artist may have created a piece of art and may be interested in selling this or at least to get recognized as the creator, such as Thingiverse users who post a design under the Creative Commons license. But with digital designs, it is easy to copy, and a Thingiverse user discovered that another user had downloaded the designs and was selling 3D printed objects of the design on eBay in violation of the Creative Commons license. eBay eventually removed them.[8] One way for designers to avoid this situation is to not share the design itself. For instance, when a designer works with a company such as BuildBee, then a G-code that is compatible with a specific 3D printer and material is shared with a customer instead of the actual design. This makes the design less susceptible for illegal copying.

What these examples demonstrate is that AM and related technologies such as scanning have introduced new vulnerabilities in IP protection and liability where it isn't always clear what is legally protected and who is liable, and even if something is legally protected, it may not be (financially) possible to enforce it.

A special case is that of 3D printed guns. This has received particular attention in the US, which in recent years has been struggling politically with gun ownership, and AM adds another dimension. The issue with guns is that the industry is regulated. That means guns have identification numbers so that they can be traced and not everybody is legally allowed to make guns (Johnson, 2013). To be more precise, what matters is certain parts of a gun, while many other parts of a gun can easily be bought. A problem is posed by guns that are not registered and thus not traceable (i.e., ghost guns). Ghost guns are not unique to AM, and the ability to make ghost guns is not dependent upon AM. Rather, in addition to AM technology, several other TM tools such as CNC machines have now become available in desktop versions and can be used to make ghost guns. An example is a reporter who used a TM process to produce a ghost gun AR15 rifle.[9] Nevertheless, the thought is that, with AM, the process of producing a ghost gun becomes easy for prosumers. Desktop AM involves mostly plastics, so an additional concern is that 3D printed guns may go undetected through security checks, although technology is already developed to detect non-metal weapons. Furthermore, while examples exist of plastic guns produced with AM, these aren't exactly good (limited capability) or even safe to use.

Regardless of the current effectiveness of 3DP guns, governments have been struggling to keep up with the technology and to establish legal frameworks. A prime example comes from the US, where the government is engaged in a legal battle with Defense Distributed. Defense Distributed is promoting the rights to make guns, has shared files for a 3D printed gun, and sells a small CNC machine that enables the manufacturing of the critical parts of a gun. The legal issues revolve around three areas: (a) sharing files for 3D printed guns, (b) producing ghost guns, and (c) owning 3D printed or ghost guns. These issues and concerns also apply in other countries such as Australia,[10] Canada and Spain,[11] The Netherlands[12] and Europe in general,[13] and in recent years, 3D printed gun-related arrests have been increasing.[14]

In early 2022, the Joe Biden administration proposed new rules dealing with the identification of guns and gun parts. Plus, in some cities in the US, a gun buyback scheme was introduced where owners of ghost guns could turn them in to the local authorities so that ghost guns were taken out of the market. However, while this regulation deals with ghost guns, the rules are not specific to people making their own guns through AM. Furthermore, the gun buyback may have created a market for more 3D printed ghost guns. For instance, in one story, a man in New York stated that because of the program he decided to 3D print guns and then sold them in a gun buyback situation for $21,000.[15] In another story, a man from Houston sold dozens of 3D printed guns at the city's first gun buyback for $3,000. They cost $3 to make and he sold them for $50 each. Apparently, this was not to make a profit but rather to send a signal to politicians about spending $1 million tax dollars on "something that has no evidence of any effect on crime".[16]

Legality of processes

Numerous AM developments have taken place since the first AM patent was granted in 1984, as was shown in Chapter 11. Seven main types of AM are recognized, but for each of these types, there are underlying processes that have differences. Companies put R&D into the development of these processes in order to gain a competitive advantage and typically apply for patents so that this advantage is protected. This gives companies time to capitalize on the advantage by, for example, selling AM machines that are based on their patented processes. Sometimes, IP disputes arise about the protection of these AM processes or components of AM machines and the raw materials used in 3DP. Notable companies that have been involved in patent litigation on AM processes are 3D Systems, DuPont, EOS, Envisiontec and Stratasys (Bechtold, 2015). This is usually a complex matter, such as the lawsuit filed by Continuous Composites against Markforged in July 2021 for infringement of its patent on simultaneously extruding two or more materials while being encased within a liquid material and where the process is for curing liquid material after extrusion that yields parts with enhanced capabilities, not possible with one material.[17] Sometimes, counter-lawsuits are filed. Desktop Metal and Markforged have been filing lawsuits against each other since 2017. Another example is Organovo and Cellink which have been involved in infringement of bioprinting patents since at least 2019. Sometimes, agreements are reached, such as in March 2022, when Organovo and Cellink agreed on a license. What these examples show is that with the new technology and high stakes, companies end up in litigation processes that are complicated and take a long time to solve. This is likely going to continue for the near future.

Safety

Among the dangers with new technologies are that it takes time to learn about safety issues and that safety risks can initially be poorly understood. It was exciting when technologically commercial aircraft moved to jets in the early 1950s, but after just a couple of years, BOAC, the airline that started flying the De Havilland Comet jet aircraft, experienced two crashes in a short span of time. It turned out that jets, which were flying at higher altitudes than previous aircraft, experienced repeated pressurization causing expansion and contraction of the metal frame and the Comet was unable to withstand this (Gero, 2010). This safety risk was not understood when the Comet was designed.

Aside from the known risks of the use of toxic materials such as chemicals in VPP and metal powders in PBF processes, AM has similar issues due to uncertainty with a new type of technology. Examples of accidents are that somebody's hair got caught in a 3D printer,[18] an unattended 3D printer almost burned down a house,[19] several 3DP materials are not always safe for food,[20] the emissions with 3DP can be toxic (Wright, 2022), and the use of hairspray to make objects stick to the printbed led to an explosion that killed a teenage boy.[21] Sometimes, lawsuits occur, such as over a death caused by a fire due to an overheated 3D printer.[22] There are continued developments to improve on these safety and legal risks, such as the development of non-toxic resins,[23] structural coloration instead of dyeing (which is more toxic for color printing),[24] the BlazeCut™ automatic fire suppression system to prevent 3D printer fires,[25] and enclosures to prevent exposure to toxic fumes,[26] and regulatory bodies such as the FDA are developing procedures to indicate safety of materials for food and medical applications (Beitler et al., 2022).

Scams involving AM

Another consequence of high technology is that the first group of adopters are taking risks because the technology is not fully understood. Part of this is investing in 3DP-related companies. As with any new company, there is always a risk with this type of investing because despite initial positive news and developments, a company may end up not being successful. ONO[27] is a case in point. The idea was to have a VPP 3D printer that was using a smartphone as the source of light. In other words, this was a type of box that could be put on top of a smartphone and 3D print small items. In 2016, the company started a Kickstarter campaign to raise money and was able to raise over $2.3 million (well above the target $80,000) with the support of over 16,000 people. Unfortunately, things did not go well. It should be noted, though, that the failure for things to go smoothly isn't strange when dealing with new technology. Formlabs, now well established, had a Kickstarter campaign in 2012, when it raised almost $3 million from over 2,000 investors, but subsequently there were delays with the delivery and some users had problems and were dissatisfied, but overall Formlabs was able to overcome these problems. In contrast, ONO was not able to overcome its challenges and may have been too optimistic or misleading in its communication to investors. In 2021, it essentially gave up.[28]

There are situations, unlike this type of unintentional failure, where people are intentionally misled. An example is a fraudulent Kickstarter campaign by Next Dynamics.[29] This campaign was suspended after people became suspicious, and there were claims that a demonstration video was fake. Rather than adding materials, the makers of the video sliced an already formed product (an egg) and then reversed the recording to make it seem as if the object was being formed layer-by-layer.[30] Other examples of fraudulent campaigns are from NEA 3D[31] and XEOS, which apparently came from the same imposter.[32] Over time, when the technology becomes more established and better known so that it will be more broadly understood what is or is not technologically possible, it is likely that these types of scams will occur less frequently.

Notes

1 https://3dprintingindustry.com/news/youtuber-combines-speed-and-style-with-custom-3d-printed-crocs-202700/
2 See https://www.internationalpropertyrightsindex.org/
3 https://www.plasticstoday.com/medical/updated-oem-threatens-sue-startup-saved-lives-3d-printing-medical-valves
4 https://www.thetmca.com/katy-perry-last-years-super-bowl-and-a-shark-with-a-mark/
5 https://abovethelaw.com/2021/03/imminent-win-for-the-public-domain-court-likely-to-compel-musee-rodin-to-release-its-3d-scans-of-sculptors-works-for-free/
6 https://www.fabbaloo.com/news/surgery-with-bioartificial-organs-legal-troubles-and-3d-printing
7 https://www.fda.gov/medical-devices/3d-printing-medical-devices/3d-printing-medical-devices-point-care-discussion-paper
8 But the seller then sued; see https://all3dp.com/just-3d-print-brings-new-lawsuit-against-3dr-holdings/
9 https://www.wired.com/2015/06/i-made-an-untraceable-ar-15-ghost-gun/ there is also a YouTube video about this https://www.youtube.com/watch?v=BTojV_NqWCA
10 https://www.3dnatives.com/en/australian-police-find-3d-printed-gun-140620224/
11 https://3dprint.com/293711/law-enforcement-in-australia-spain-u-s-and-canada-continue-targeting-3d-printed-guns/
12 https://3dprint.com/291819/dutch-police-confirm-increase-in-3d-printed-guns/

13 https://www.3dnatives.com/en/europol-concerned-3d-printed-weapons-08062022/
14 https://3dprint.com/291684/3d-printed-gun-arrests-tripled-in-less-than-two-years-3dprint-com-investigates/
15 https://3dprint.com/294903/man-gets-21000-for-3d-printed-guns-in-ny-what-does-it-mean-for-gun-buybacks/
16 https://3dprint.com/293221/3d-printed-gun-arrests-raids-and-a-man-who-got-3000-at-a-gun-buyback/
17 https://3dprintingindustry.com/news/markforged-challenges-continuous-composites-3d-printing-patent-lawsuit-210143/
18 https://www.fabbaloo.com/2020/01/more-on-the-3d-printer-hair-incident
19 https://all3dp.com/fire-safety-unattended-3d-printer-nearly-burns-house-down/
20 https://3dinsider.com/food-safe-3d-printed-items/
21 https://www.3ders.org/articles/20161103-explosion-caused-by-3d-printer-and-hairspray-kills-teenage-boy.html
22 https://www.3dnatives.com/en/alibaba-and-tronxy-3d-sued-140620225/#!
23 https://www.fabbaloo.com/news/it-turns-out-there-are-non-toxic-3d-printer-resins
24 https://3dprintingindustry.com/news/scientists-come-up-with-new-method-of-creating-safer-colored-3d-printed-parts-210046/
25 https://www.3dupfitters.com/blogs/news/lets-talk-about-3d-printer-fire-safety-will-this-thing-burn-down-my-house
26 https://all3dp.com/2/3d-printer-enclosure-5-cheap-diy-options/
27 Initially OLO but rebranded due to claims from another company, see https://www.kickstarter.com/projects/olo3d/olo-the-first-ever-smartphone-3d-printer/posts/1642295
28 https://3dprintingindustry.com/news/ono-pulls-plug-on-smartphone-3d-printer-leaving-backers-empty-handed-192889/
29 https://3dprintingindustry.com/news/next-dynamics-multi-material-nexd1-3d-printer-passes-half-million-mark-kickstarter-103680/
30 https://3dprintingindustry.com/news/nexd1-kickstarter-suspended-amidst-controversy-104318/
31 https://www.kickstarter.com/dmca/nea-canceled-submitted-by-nea-3d-ltd
32 https://3dprintingindustry.com/news/kickstarter-fraud-launches-another-3d-printer-campaign-48284/

14 Conclusion

The world is going through a fourth industrial revolution, known as Industry 4.0, which relates to cyber-physical systems. These systems have several core underlying technologies such as the Internet of Things and autonomous robots. AM is also one of the key technologies that make Industry 4.0 possible. This chapter will draw main conclusions by looking at commonalities that can be detected across the previous chapters, particularly Chapters 6 to 10.

AM was initially used for prototyping was then used for tooling and is now applied for limited production of parts and products. Across Chapters 6 to 10, a large variety of these types of applications were shown. The common element is that many of these applications of AM have to do with small-volume production. This makes sense in light of Figures 3.6 and 3.7, which showed that for small volumes, AM is often a better option than TM in terms of cost and time. Therefore, companies may find that if they are already producing a (large) variety of low-volume items with TM, then AM might be a better alternative. A prime example of this is shoes. It also applies to spare parts, especially for older products where the demand for each part is low but there is demand for a large variety. In this instance, there is also a trade-off between using TM with large production runs and having inventory costs or using AM on demand. What this means is that AM should be considered as one method of production in addition to other production methods. Whether AM is the best alternative depends upon whether the circumstances have the right fit. AM is not the best option, or even feasible, for every situation. If there is demand for a limited volume, then AM might be a good option. This is what occurs when a prototype is needed. It also occurs with a single metal mold (tool) and with entrepreneurial product development and prosumer activities where small quantities are produced.

Another major reason why AM is adopted is that it allows customization. Customization is typically related to unique products for which potentially a price premium can be asked. This uniqueness also connects to low volume. Prime examples of this are customized jewelry and applications in the medical field that require patient-specific items such as prosthesis. Sometimes this also means that small-volume customized products now can be produced that previously were not financially feasible, such as a surgery model.

DOI: 10.4324/9781003399100-17

Furthermore, more expensive types of AM such as MJT can be attractive, as shown for clothes in Figure 6.2. Customization is also related to product complexity. When products have complex shapes and require a lot of parts and tools, then AM becomes more appealing. Many examples of this are found in aerospace.

From Chapters 6 to 10, it follows that there are instances where the adoption of industrial AM may be in the innovators stage, such as with food. There are also indications that the early majority stage, such as for hearing aids, may already have been reached. However, most of the AM adoption is somewhere in the stage of the early adopters. There are two main challenges for the further spread of AM. First, there is the challenge of crossing the chasm. The early majority adopters are distinctly different from the innovators and early adopters. The latter two are more willing to deal with technology uncertainty and look for potential competitive advantages. The early majority is much more pragmatic and looks for proven technology, how it fits in their organization and how it affects productivity. Many manufacturers of AM equipment thus are working on demonstrating that the AM technology has a sufficient level of maturity, while the legal system is catching up with the newness of AM technology, and new applications that have potential impact on competitive strategy, such as in the area of sustainability, still frequently occur.

The second main challenge, related to this, is that industry is looking for scaling things and that means having larger production runs. However, as was explained above, a key characteristic of AM is that it is advantageous for smaller production volumes. Logically then, what is observed for AM technology developments (Chapter 11) are developments that extend capabilities by making processes faster, cheaper – through requiring fewer resources and thereby reducing cost (e.g., involving less post-processing) – and higher in quality (such as through better and stronger materials which often cannot be used in TM). These developments extend the number of situations in which AM can be applied. A great example of that is the large number of developments in aerospace related to AM. Some of these developments also improve the volume for which AM is beneficial compared with TM. When going back to Figure 3.7, if AM becomes faster then the AM line becomes steeper and, therefore, the point at which the AM and TM lines intersect moves to the right, meaning that AM becomes competitive for higher volumes.

On the prosumer side, there is a heavy emphasis on the use of MEX as the main AM technology; however, it is used in many different ways. Chapters 6 to 10 provide many examples of individuals who have experimented with AM. The people involved are technology enthusiasts who sometimes have design backgrounds and an ability to apply technical knowledge, such as Julia Daviy and Danit Peleg with clothes, Austen Hartley and Stephan Henrich with shoes, and Chloé Rutzerveld and Daniel Wilkens with food. They often create new things that don't necessarily replace an existing market but have the potential to lead to new markets. These characteristics fit with that of the first adopter group: the innovators. It is not clear whether the early adopter stage has been reached for prosumers, but it is extremely unlikely that the early majority has been reached. The early majority prosumer is pragmatic and would look for established technology that works. This is more like push-button technology. Although significant progress has been made in recent years for desktop AM, it does not yet have that level of maturity.

Based on this discussion for the industrial and prosumer side, it is clear that AM is generally still in an early stage of diffusion which means that it is highly likely that in the near-term, AM is going to be adopted much more. A consequence of this will be that there will be many employment opportunities related to AM. AM companies that grow will need a variety of people, including for business functions such as marketing and accounting. They will also need people who can handle the technology. People with design skills will become especially important because, as discussed in Chapter 2, AM works differently than TM. This affects how products are designed. The true potential of AM may not be reached until a new generation of designers comes into the workforce that have grown up with AM technology and, because of this, have a much different mindset about how to design.

Theoretically, AM is more or less location-independent and can be applied everywhere, even in space, as long as items like raw materials are available. To some degree, this is observed with countries, including less developed ones, across the globe engaging in AM. On the African continent, countries such as Ghana, Kenya and Nigeria are active in AM while South Africa has been the most active with companies such as Solid Edge Technologies. AM can be tailored to circumstances such as making affordable tools for farmers in Africa.[1] Latin American countries are a bit behind with AM, but Argentina and Brazil are getting involved, and in Columbia, AM is used for construction. Some countries even have explicit strategies for increasing AM. The United Arab Emirates has an emphasis on AM construction. The US has America Makes, a public–private partnership, and the AM Forward initiative, which aim to increase AM and its adoption by industry. India has a National Strategy on Additive Manufacturing, whose goal is to capture 5% of the global market share in AM by 2025. Denmark has Dansk AM Hub, a business foundation with a vision to make Denmark a world leader in the use of AM, and Russia plans to have 180 centers for additive technologies by 2030.[2]

An advantage of the theoretical location independence is the opportunity to produce closer to the market and avoid the logistics cost of shipping goods from centralized locations. The variety of situations presented in Chapters 6 to 10 shows that, to some extent, this is indeed the case (e.g., in situations where someone, or some company, adopts AM and starts to manufacture items that previously were not produced in that particular location or when 3DP on a ship or in the ISS). However, as was also shown in Chapters 6 to 10, in regard to where in the value chain AM gets used, there are often still centralized locations involved. A reason for this is that the current AM needs for companies can be too small to warrant having duplicative and expensive equipment in multiple locations. Furthermore, sometimes centralization happens when outside parties, such as Sculpteo, are involved. A reason for outsourcing AM is that the current AM needs for companies can be too small to warrant owning expensive AM equipment. Companies, such as Sculpteo, with centralized AM facilities produce for a variety of client companies and thus can combine many small orders to reach a high degree of utilization for its AM equipment.

Another location-related observation is that some AM seems to be concentrated in certain geographical areas. Several companies in Israel, such as Plantish, Redefine Meat, SavorEat and Steakholder Foods, deal with AM of food, and The Netherlands has a broad range of AM companies, such as Builder 3D, Felix and Ultimaker (AM machines), CyBe Construction (construction), 3D Hubs (AM services), ColorFabb (filament) and 3devo (recycling plastic for AM filament). These might be indications of the initial development of innovative AM clusters (see Porter and Stern, 2001). The future will tell whether clusters will develop in specific locations that specialize in specific areas related to AM.

Notes

1 https://3dprintingindustry.com/news/researchers-3d-print-recycled-plastic-into-affordable-tools-for-african-farmers-196015/
2 https://additiv-tech.ru/en/news/new-association-development-additive-technologies-aims-advance-3d-printing-russia.html

References

Abele, E., Meyer, T., Näher, U., Strube, G. and Sykes, R. (Eds.) 2008. *Global production: A handbook for strategy and implementation.* Berlin: Springer-Verlag.

Akmal, J.S., Salmi, M., Björkstrand, R., Partanen, J. and Holmström, J. 2022. "Switchover to industrial additive manufacturing: dynamic decision-making for problematic spare parts." *International Journal of Operations & Production Management*, Vol. 42, No. 13, pp. 358–384.

Alfonso, S., Gesto, M. and Sadoul, B. 2021. "Temperature increase and its effects on fish stress physiology in the context of global warming." *Journal of Fish Biology*, Vol. 98, No. 6, pp. 1496–1508.

Alkhalidi, A. and Hatuqay, D. 2020. "Energy efficient 3D printed buildings: material and techniques selection worldwide study." *Journal of Building Engineering*, Vol. 30, article 101286.

Al-Mazrouei, N., Al-Marzouqi, A.H. and Ahmed, W. 2022. "Characterization and sustainability potential of recycling 3D-printed nylon composite wastes." *Sustainability*, Vol. 14, pp. 10458.

Anderson, C. 2012. *Makers: The next industrial revolution.* New York: Crown Business.

Arvanitidi, E., Drosos, C., Theocharis, E. and Papoutsidakis, M. 2019. "3D printing and education." *International Journal of Computer Applications*, Vol. 177, No. 24, pp. 55–59.

Ashraf, H., Meer, B., Naz, R., Saeed, A. Sadia, H., Sajid, U., Nisar, K., Aslam, Z. and Anwar, P. 2018. "3D-bioprinting: a stepping stone towards enhanced medical approaches." *Advancements in Life Sciences*, Vol. 5, No. 4, pp. 143–153.

Assante, D., Cennamo, G.M. and Placidi, L., 2020, "3D Printing in Education: an European perspective". *2020 IEEE Global Engineering Education Conference (EDUCON)*, pp. 1133–1138.

Atzeni, E., Iuliano, L., Minetola, P. and Salmi, A. 2010. "Redesign and cost estimation or rapid manufactured plastic parts." *Rapid Prototyping Journal*, Vol. 16, No. 5, pp. 308–317.

Atzeni, E. and Salmi, A. 2012. "Economics of additive manufacturing for end-usable metal parts". *International Journal Advanced Manufacturing Technology*, Vol. 62, pp. 1147–1155.

Avrutis, D., Nazari, A. and Sanjayan, J.G. 2019. "Industrial adoption of 3D concrete printing in the Australian market: potentials and challenges." In: Sanjayan, J.G., Nazari, A. and Nematollahi, B. (Eds.) *3D concrete printing technology: Construction and building applications*. Oxford: Butterworth-Heinemann.

Baiano, A. 2020. "3D printed foods: a comprehensive review on technologies, nutritional value, safety, consumer attitude, regulatory framework, and economic and sustainability issues." *Food Reviews International*. DOI: 10.1080/87559120.2020.1762091.

Ballardini, R.M., Ituarte, I.F. and Pei, E. 2018. "Printing spare parts through additive manufacturing: legal and digital business challenges." *Journal of Manufacturing Technology Management*, Vol. 29, No. 6, pp. 958–982.

Barbosa, F. et al. 2017. *Reinventing construction: a route to higher productivity*. McKinsey Global Institute.

Basak, S., Baumers, M., Holweg, M., Hague, R. and Tuck, C. 2022. "Reducing production losses in additive manufacturing using overall equipment effectiveness." *Additive Manufacturing*, Vol. 56, pp. 102904.

Bechtold, S. 2015. "3D printing and the intellectual property system." Economic Research Working Paper No. 28, World Intellectual Property Organization.

Beck, J.M. and Jacobson, M.D. 2017. "3D printing: what could happen to products liability when users (and everyone else in between) become manufacturers." *Minnesota Journal of Law, Science & Technology*, Vol. 18, No. 1, pp. 143–205.

Beer, A. 2018. "The closure of the Australian car manufacturing industry: redundancy, policy and community impacts." *Australian Geographer*, Vol. 49, No. 3, pp. 419–438.

Beitler, B.G., Abraham, P.F., Glennon, A.R., Tommasini, S.M., Lattanza, L.L., Morris, J.M. and Wiznia, D.H. 2022. "Interpretation of regulatory factors for 3D printing at hospitals and medical centers, or at point of care." *3D Printing in Medicine*, Vol. 8, No. 1, pp. 1–7.

Bender, H.O. 1989. "Marketing technologically advanced products." *European Management Journal*, Vol. 7, No. 2, pp. 159–165.

Benstead, A.V., Stevenson, M. and Hendry, L.C. 2017. "Why and how do firms reshore? A contingency-based conceptual framework." *Operations Management Research*, Vol. 10, No. 3/4, pp. 85–103.

Bernstein, W.J. 2004. *The birth of plenty: How the prosperity of the modern world was created*. New York: McGraw-Hill.

Bezek, L.B., Pan, J., Harb, C., Zawaski, C.E., Molla, B., Kubalak, J.R., Marr, L.C. and Williams, C.B. 2021. "Additively manufactured respirators: quantifying particle transmission and identifying system-level challenges for improving filtration efficiency." *Journal of Manufacturing Systems*, Vol. 60, pp. 762–773.

Bhattacharyya, S.S., Kulkarni, O. and Mishra, A. 2022. "Study of emerging avenues in supply chain resilience; the case of integration of additive manufacturing with spare parts procurement." *Benchmarking: An International Journal*, (ahead-of-print).

Bishop, E.S., Mostafa, S., Pakvasa, M., Luu, H.H., Lee, M.J., Wolf, J.M., Ameer, G.A., He, T.C. and Reid, R.R. 2017. "3-D bioprinting technologies in tissue engineering and regenerative medicine: current and future trends." *Genes & Diseases*, Vol. 4, pp. 185–195.

Blakey-Milner, B., Gradl, P., Snedden, G., Brooks, M., Pitot, J., Lopez, E., Leary, M., Berto, F. and du Plessis, A. 2021. "Metal additive manufacturing in aerospace: review." *Materials & Design*, Vol. 209.

Bower, J.L. and Christensen, C.M., 1995. "Disruptive technologies: catching the wave." *Harvard Business Review*, Vol. 73, No. 1, pp. 43–53.

Brander, K.M., 2018. "Climate change not to blame for cod population decline." *Nature Sustainability*, Vol. 1, No. 6, pp. 262–264.

Brean, D.H. 2013. "Asserting patents to combat infringement via 3D printing: it's no "use"." *Fordham Intellectual Property, Media & Entertainment Law Journal*, Vol. 23, pp. 771–814.

Brem, A., Nylund, P.A. and Schuster, G. 2016. "Innovation and de facto standardization: the influence of dominant design on innovative performance, radical innovation, and process innovation." *Technovation*, Vol. 50–51, pp. 79–88.

Brown, A.S. 2019. "3D printed figures take the cake." *Mechanical Engineering*, May, pp. 64.

Campoy-Muñoz, P., Cardenete, M.A. and Delgado, M.C. 2017. "Economic impact assessment of food waste reduction on European countries through social accounting matrices." *Resources, Conservation and Recycling*, Vol. 122, pp. 202–209.

Chen, C., van Groenigen, K.J., Yang, H., Hungate, B.A., Yang, B., Tian, Y., Chen, J., Dong, W., Huang, S., Deng, A. and Jiang, Y. 2020a. "Global warming and shifts in cropping systems together reduce China's rice production." *Global Food Security*, Vol. 24, p. 100359.

Chen, G., Xu, Y., Kwok, P.C.L. and Kang, L. 2020b. "Pharmaceutical applications of 3D printing." *Additive Manufacturing*, Vol. 34.

Christensen, C.M. 2011. *The innovator's dilemma*. New York: Harper Business.

Christensen, C.M., Raynor, M. and McDonald, R. 2015. "What is disruptive innovation?" *Harvard Business Review*, Vol. 93, No. 12, pp. 44–53.

Chua, C.K. and Leong, K.F. 2017. *3D printing and additive manufacturing: Principles and applications*. Singapore: World Scientific Publishing Co.

Cooper, F. 2016. "Sintering and additive manufacturing: the new paradigm for the jewellery manufacturer." *Progress in Additive Manufacturing*, Vol. 1, pp. 29–43.

Cotteleer, M.J. 2014. "3D opportunity for production, additive manufacturing makes its (business) case." *Deloitte Review*, Vol. 15, pp. 146–161.

Coykendall, J., Cotteleer, M., Holdowsky, J. and Mahto, M. 2014. *3D opportunity in aerospace and defense: Additive manufacturing takes flight*. Deloitte University Press.

Danielmeyer, H.G. and Takeda, Y. 1999. *The company of the future: Markets, tools, and strategies*. Berlin: Springer-Verlag.

Davis, F.D. 1989. "Perceived usefulness, perceived ease of use, and user acceptance of information technology." *MIS Quarterly*, Vol. 13, No. 3, pp. 319–340.

Dawood, A., Marti Marti, B., Sauret-Jackson, V. and Darwood, A. 2015. "3D printing in dentistry." *British Dental Journal*, Vol. 219, No. 11, pp. 521–523.

Dedehayir, O. and Steinert, M. 2016. "The hype cycle model: a review and future directions." *Technological Forecasting & Social Change*, Vol. 108, pp. 28–41.

Demlehner, Q., Schoerner, D. and Laumer, S. 2021. "How can artificial intelligence enhance car manufacturing? A Delphi study-based identification and assessment of general use cases." *International Journal of Information Management*, Vol. 58.

Deng, S., Wang, R.Z. and Dai, Y.J. 2014. "How to evaluate performance of net zero energy buildings – a literature research." *Energy*, Vol. 71, pp. 1–16.

Depoorter, B. 2014. "Intellectual property infringements & 3D printing: decentralized piracy." *Hastings Law Journal*, Vol. 65, pp. 1483–1594.

Devarajan, B., LakshmiNarasimhan, R., Venkateswaran, B., Mavinkere Rangappa, S. and Siengchin, S. 2022. "Additive manufacturing of jute fiber reinforced polymer composites: a concise review of material forms and methods." *Polymer Composites*, Vol. 43, No. 10, pp. 6735–6748.

Dicken, P. 2011. *Global shift: Mapping the changing contours of the world economy*. New York. The Guilford Press.

Dilberoglu, U.M., Gharehpapagh, B., Yaman, U. and Dolen, M. 2017. "The role of additive manufacturing in the era of industry 4.0." *Procedia Manufacturing*, Vol. 11, pp. 545–554.

Ding, T., Xiao, J., Zou, S. and Wang, Y. 2020. "Hardened properties of layered 3D printed concrete with recycled sand." *Cement and Concrete Composites*, Vol. 113, article 103724.

Donahue, M.Z. 2015. "Doctors can study 3D printed models of your organs before surgery." *Smithsonian Magazine*. https://www.smithsonianmag.com/innovation/doctors-can-study-3d-printed-models-your-organs-surgery-180955622/

Dosi, G. and Galambos, L. (Eds.) 2013. *The third industrial revolution in global business*. Cambridge: Cambridge University Press.

Duran, D.C., Gogan, L.M., Artene, A. and Duran, V. 2015. "The components of sustainable development – a possible approach." *Procedia Economics and Finance*, Vol. 26, pp. 806–811.

Elsayed, H., Gobbin, F., Picicco, M., Italiano, A. and Colombo, P. 2022. "Additive manufacturing of inorganic components using a geopolymer and binder jetting." *Additive Manufacturing*, Vol. 56, p. 102909.

Emelogu, A., Maruffuzzaman, M., Thompson, S.M., Shamsaei, N. and Bian, L. 2016. "Additive manufacturing of biomedical implants: a feasibility assessment via supply-chain cost analysis." *Additive Manufacturing*, Vol. 11, pp. 97–113.

Eriksson, S. and Steenhuis, H.J. (Eds.) 2016. *The global commercial aviation industry*. London: Routledge.

Fan, F., Li, S., Huang, W. and Ding, J. 2022. "Structural characterization and fluidness analysis of lactose/whey protein isolate composite hydrocolloids as printing material for 3D printing." *Food Research International*, Vol. 195.

Feng, P., Meng, X., Chen, J.F. and Ye, L. 2019. "Mechanical properties of structures 3D-printed with cementitious powders." In: Sanjayan, J.G., Nazari, A. and Nematollahi, B. (Eds.) *3D concrete printing technology, Construction and building applications*. Oxford: Butterworth-Heinemann.

Ferdows, K. 1997. "Making the most of foreign factories." *Harvard Business Review*, Vol. 75, pp. 73–88.

Ferguson, N. 2011. *Civilization: The west and the rest*. New York: The Penguin Press.

Fleury, A., Gregory, M. and Bennett, D. 2007. "The future of manufacturing." *Journal of Manufacturing Technology Management*, Vol. 18, No. 8.

Ford, S. and Minshall, T. 2019. "Invited review article: where and how 3D printing is used in teaching and education". *Additive Manufacturing*, Vol. 25, pp. 131–150.

Fransen, M.F., et al. 2021. "Bioprinting of kidney in vitro models: cells, biomaterials, and manufacturing techniques." *Essays in Biochemistry*, Vol. 65, No. 3, pp. 587–602.

Fratocchi, L. 2018. "Additive manufacturing technologies as a reshoring enabler: a why, where and how approach." *World Review of Intermodal Transportation Research*, vol. 7, No. 3, pp. 264–293.

Froes, F. and Boyer, R. (Eds.) 2019. *Additive manufacturing for the aerospace industry*. Amsterdam: Elsevier.

Garg, A., Yerneni, S.S., Campbell, P., LeDuc, P.R. and Ozdoganlar, O.B. 2022. "Freeform 3D ice printing (3D-ICE) at the micro scale." *Advanced Science*, Vol. 9, No. 27, pp. 2201566.

Gasman, L. 2019. "Additive aerospace considered as a business." In Froes, F. and Boyer, R. (Eds.) *Additive manufacturing for the aerospace industry*. Amsterdam: Elsevier.

Gereffi, G. 1999. "International trade and industrial upgrading in the apparel commodity chain." *Journal of International Economics*, Vol. 48, pp. 37–70.

Gero, D. 2010. *Aviation disasters: The world's major civil airliner crashes since 1950*, 5th edition. Gloucestershire (UK): The History Press.

Gisario, A., Kazarian, M., Martina, F. and Mehrpouya, M. 2019. "Metal additive manufacturing in the commercial aviation industry: a review." *Journal of Manufacturing Systems*, Vol. 53, pp. 124–149.

Glavič, P. and Lukman, R. 2007. "Review of sustainability terms and their definitions." *Journal of Cleaner Production*, Vol. 15, pp. 1875–1885.

Godoi, F.C., Bhandari, B.R., Prakash, S. and Zhang, M. (Eds.) 2019a. *Fundamentals of 3D food printing and applications*. London: Academic Press.

Godoi, F.C., Bhandari, B.R., Prakash, S. and Zhang, M. 2019b. "An introduction to the principles of 3D food printing." In: Godoi, F.C., Bhandari, B.R., Prakash, S. and Zhang, M. (Eds.) *Fundamentals of 3D food printing and applications*. London: Academic Press.

Goosen, M.F.A. 2012. "Environmental management and sustainable development." *Procedia Engineering*, Vol. 33, pp. 6–13.

Gray, J.V., Esenduran, G., Rungtusanatham, M.J. and Skowronski, K. 2017. "Why in the world did they reshore? Examining small to medium-sized manufacturer decisions." *Journal of Operations Management*, Vols 49–51, pp. 37–51.

Gu, Z., Fu, J., Lin, H. and He, Y. 2020. "Development of 3D bioprinting: from printing methods to biomedical applications." *Asian Journal of Pharmaceutical Sciences*, Vol. 15, pp. 529–557.

Gupta, M.C. 1995. "Environmental management and its impact on the operations function." *International Journal of Operations & Production Management*, Vol. 15, No. 8, pp. 34–51.

Habibi, M., Foroughi, S., Karamzadeh, V. and Packirisamy, M. 2022. "Direct sound printing." *Nature Communications*, Vol. 13, No. 1, pp. 1800.

Hahn, V., Rietz, P., Hermann, F., Müller, P., Barner-Kowollik, C., Schlöder, T., Wenzel, W., Blasco, E. and Wegener, M. 2022. "Light-sheet 3D microprinting via two-colour two-step absorption." *Nature Photonics*, Vol. 16, No. 11, pp. 784–791.

Hambach, M., Rutzen, M. and Volkmer, R. 2019. "Properties of 3D-printed fiber-reinforced Portland cement paste." In: Sanjayan, J.G., Nazari, A. and Nematollahi, B. (Eds.) *3D concrete printing technology, Construction and building applications*. Oxford: Butterworth-Heinemann.

Hasegawa, T., Havlik, P., Frank, S., Palazzo, A. and Valin, H. 2019. "Tracking food consumption inequality to fight hunger without pressuring the environment." *Nature Sustainability*, Vol. 2, pp. 826–833.

Hayes, R.H. and Wheelwright, S.C. 1984. *Restoring our competitive edge: Competing through manufacturing*. New York: John Wiley & Sons.

He, Y., Zhang, Y., Zhang, C. and Zhou, H. 2020. "Energy-saving potential of 3D printed concrete building with integrated living wall." *Energy & Buildings*, Vol. 222, article 110110.

Heard, T.A. 1999. "The role of stingless bees in crop pollination." *Annual Review of Entomology*, Vol. 44, No. 1, pp. 183–206.

Heinen, J.J. and Hoberg, K. 2019. "Assessing the potential of additive manufacturing for the provision of spare parts." *Journal of Operations Management*, Vol. 65, No. 8, pp. 810–826.

Hernandez, P. and Kenny, P. 2010. "From net energy to zero energy buildings: defining life cycle zero energy buildings (LC-ZEB)." *Energy and Buildings*, Vol. 42, pp. 815–821.

Hill, T. 2000. *Manufacturing strategy*, 3rd edition. Boston: Irwin McGraw-Hill.

Hohn, M.M. and Durach, C.F. 2021. "Additive manufacturing in the apparel supply chain – impact on supply chain governance and social sustainability." *International Journal of Operations & Production Management*, Vol. 41, No. 7, pp. 1035–1059.

Holmström, J., Partanen, J., Tuomi, J. and Walter, M. 2010. "Rapid manufacturing in the spare parts supply chain." *Journal of Manufacturing Technology Management*, Vol. 21, No. 6, pp. 687–697.

Holt, C., Edwards, L., Keyte, L., Moghaddam, F. and Townsend, B. 2019. "Construction 3D printing." In: Sanjayan, J.G., Nazari, A. and Nematollahi, B. (Eds.) *3D concrete printing technology: Construction and building applications*. Oxford: Butterworth-Heinemann.

Hooton, C.A. 2018. "Defining tech: an examination of how the 'technology' economy is Measured." *Journal of NBICT*, Vol. 1, pp. 101–120.

Hornick, J. 2015. *3D printing will rock the world*. North Charleston, SC: CreateSpace Independent Publishing Platform.

Huang, I. and Park, J.H. 2018. "The international competitive relationships of three leading countries in the global shipbuilding industry." *The Korean Journal of Policy Studies*, Vol. 33, No. 3, pp. 73–91.

Jabaly, P. 2010. "IP litigation or ADR: costing out the decision." *Journal of Intellectual Property Law & Practice*, Vol. 5, No. 10, pp. 730–735.

Jacometti, V. 2019. "Circular economy and waste in the fashion industry." *Laws*, Vol. 8, No. 4, article 27.

Jayaprakash, S., Paasi, J., Pennanen, K., Ituarte, I.F., Lille, M., Partanen, J. and Sozer, N. 2020. "Techno-economic prospects and desirability of 3D food printing: perspectives of industrial experts, researchers and consumers." *Foods*, Vol. 9.

Jia, F., Wang, X., Mustafee, N. and Hao, L. 2016. "Investigating the feasibility of supply-chain centric business models in 3D chocolate printing: a simulation study." *Technological Forecasting & Social Change*, Vol. 102, pp. 202–213.

Johnson, J.J. 2013. "Print, lock and load: 3-D printers, creation of guns, and the potential threat to fourth amendment rights." *Illinois Journal of Law, Technology, and Policy*, Vol. 2013, No. 2, pp. 337–361.

Jose, P.A. and Christopher, P.G.V. 2018. "3D printing of pharmaceuticals – a potential technology in developing personalized medicine." *Asian Journal of Pharmaceutical Research and Development*, Vol. 6, No. 3, pp. 46–54.

Kading, B. and Straub, J. 2014. "Utilizing in-situ resources and 3D printing structures for a manned Mars mission." *Acta Astronautica*, Vol. 107, pp. 317–326.

Kamble, P.P., Chavan, S., Hodgir, R., Gote, G. and Karunakaran, K.P. 2022. "Multi-jet ice 3D printing." *Rapid Prototyping Journal*, Vol. 28, No. 6, pp. 989–1004.

Karanikas, N., Melis, D.J. and Kourousis, K.I. 2018. "The balance between safety and productivity and its relationship with human factors and safety awareness and communication in aircraft manufacturing." *Safety and Health at Work*, Vol. 9, pp. 257–264.

Karnal, M. and Rizza, G. 2019. "Design for metal additive manufacturing for aerospace applications." In: Froes, F. and Boyer, R. (Eds.) *Additive manufacturing for the aerospace industry*. Amsterdam: Elsevier.

Kazemian, A., Yuan, X., Meier, R. and Khoshnevis, B. 2019. "Performance-based testing of Portland cement concrete for construction-scale 3D printing." In: Sanjayan, J.G., Nazari, A. and Nematollahi, B. (Eds.) *3D concrete printing technology: Construction and building applications*. Oxford: Butterworth-Heinemann.

Khademhosseini, A. and Camci-Unal, G. (Eds.). 2018. *3D Bioprinting in regenerative engineering, Principles and applications*. Boca Raton: CRC Press.

Khajavi, S.H., Partanen, J. and Holmström, J. 2014. "Additive manufacturing in the spare parts supply chain." *Computers in Industry*, Vol. 65, pp. 50–63.

Khosravani, M.R. and Reinicke, T. 2020. "On the environmental impact of 3D printing technology." *Applied Materials Today*, Vol. 20, article 100689.

Kim, C. and Kim, H.J. 2019. "A study on healthcare supply chain management efficiency: using bootstrap data envelopment analysis." *Health Care Management Science*, Vol. 22, pp. 534–548.

Kim, S., Seong, H., Her, Y. and Chun, J. 2019. "A study of the development and improvement of fashion products using a FDM type 3D printer." *Fashion and Textiles*, Vol. 6, No. 9.

Kim, S. et al. 2022. "Closed-loop additive manufacturing of upcycled commodity plastic through dynamic cross-linking." *Science Advances*, Vol. 8.

Kleger, N., Fehlmann, S., Lee, S.S., Dénéréaz, C., Cihova, M., Paunović, N., Bao, Y., Leroux, J.C., Ferguson, S.J., Masania, K. and Studart, A.R., 2022. "Light-based printing of leachable salt molds for facile shaping of complex structures." *Advanced Materials*, Vol. 34, No. 32, p. 2203878.

Klein, A.M., Boreux, V., Fornoff, F., Mupepele, A.C. and Pufal, G., 2018. "Relevance of wild and managed bees for human well-being." *Current Opinion in Insect Science*, Vol. 26, pp. 82–88.

Knofius, N., van der Heijden, M.C. and Zijm, W.H.M. 2016. "Selecting parts for additive manufacturing in service logistics." *Journal of Manufacturing Technology Management*, Vol. 27, No. 7, pp. 915–931.

Koonin, S.E. 2021. *Unsettled? What climate science tells us, what it doesn't, and why it matters*. Dallas, TX: BenBella Books.

Kuznetsov, V.E., Solonin, A.N. Tavitov, A., Urzhumtsev, O. and Vakulik, A. 2020. "Increasing strength of FFF three-dimensional printed parts by influencing on temperature related parameters of the process." *Rapid Prototyping Journal*, Vol. 26, No. 1, pp. 107–121.

Kuznetsov, V.E., Tavitov, A.G., Urzhumtsev, O.D., Mikhalin, M.V. and Moiseev, A.I. 2019. "Hardware factors influencing strengths of parts obtained by fused filament fabrication." *Polymers*, Vol. 11.

Kwatra, S., Kumar, A. and Sharma, P. 2020. "A critical review of studies related to construction and computation of sustainable development indices." *Ecological Indicators*, Vol. 112, pp. 1–15.

Leach, N. 2014. "3D printing in space." *Architectural Design*, Vol. 84, No. 6, pp. 108–113.

Lee, J. 2021. "A 3D food printing process for the new normal era: a review." *Processes*, Vol. 9, article 1495.

Liker, J.K., Kamath, R.R. and Wasti, S.N. 1998. "Supplier involvement in design: a comparative survey of automotive suppliers in the USA, UK and Japan." *International Journal of Quality Science*, Vol. 3, No. 3, pp. 214–238.

Lipkowitz, G., Samuelsen, T., Hsiao, K., Lee, B., Dulay, M.T., Coates, I., Lin, H., Pan, W., Toth, G., Tate, L. and Shaqfeh, E.S. 2022. "Injection continuous liquid interface production of 3D objects." *Science Advances*, Vol. 8, No. 39.

Lipson, H. and Kurman, M. 2013. *Fabricated: The new world of 3D printing*. Indianapolis, IN: John Wiley & Sons.

Liu, Z. and Zhang, M. 2019. "3D food printing technologies and factors affecting printing precision". In: Godoi, F.C., Bhandari, B.R., Prakash, S. and Zhang, M. (Eds.) *Fundamentals of 3D food printing and applications*. London: Academic Press.

Longoni, A. and Cagliano, R. 2015. "Environmental and social sustainability priorities, their integration in operations strategies." *International Journal of Operations & Production Management*, Vol. 35, No. 2, pp. 216–245.

Luomaranta, T. and Martinsuo, M. 2022. "Additive manufacturing value chain adoption." *Journal of Manufacturing Technology Management*, Vol. 33, No. 9, pp. 40–60.

Magomedov, I.A., Khaliev, M.S.-U. and Bagov, A.M. 2020. "Agriculture and its contribution to global warming." *IOP Conferences Series: Earth and Environmental Science* Vol. 548.

Malaeb, Z., AlSakka, F. and Hamzeh, F. 2019. "3D concrete printing: machine design, mix proportioning, and mix comparison between different machine setups." In: Sanjayan, J.G., Nazari, A. and Nematollahi, B. (Eds.) *3D concrete printing technology: Construction and building applications*. Oxford: Butterworth-Heinemann.

Maldonado-Rosas, R., Tejada-Ortigoza, V., Cuan-Urquizo, E., Mendoza-Cachú, D., Morales-de La Pena, M., Alvarado-Orozco, J.M. and Campanella, O.H. 2022. "Evaluation of rheology and printability of 3D printing nutritious food with complex formulations." *Additive Manufacturing*, Vol. 58, p. 103030.

Mandrycky, C., Wang, Z., Kim, K. and Kim, D.H. 2016. "3D bioprinting for engineering complex tissues." *Biotechnology Advances*, Vol. 34, pp. 422–434.

Mantihal, S., Kobun, R. and Lee, B.B. 2020. "3D food printing of as the new way of preparing food: a review." *International Journal of Gastronomy and Food Science*, Vol. 22.

Marchese, D., Reynolds, E., Bates, M.E., Morgan, H., Clark, S.S. and Linkov, I. 2018. "Resilience and sustainability: similarities and differences in environmental management applications." *Science of the Total Environment*, Vol. 613, pp. 1275–1283.

Marchment, T., Sanjayan, J.G., Nematollahi, B. and Xia, M. 2019. "Interlayer strength of 3D printed concrete: influencing factors and method of enhancing." In: Sanjayan, J.G., Nazari, A. and Nematollahi, B. (Eds.) *3D concrete printing technology: Construction and building applications*. Oxford: Butterworth-Heinemann.

Marquis, D.G. 1969. "The anatomy of successful innovations." *Innovation*, Vol. 1, No. 7, pp. 28–37.

Marszal, A.J. et al. 2011. "Zero energy building – a review of definitions and calculation methodologies." *Energy and Buildings*, Vol. 43, pp. 971–979.

Matheson, K. et al. 2016. "Linking eelgrass decline and impacts on associated fish communities to European green crab Carcinus maenas invasion." *Marine Ecology Progress Series*, Vol. 548, pp. 31–45.

McDermott, C.M., Greis, P. and Fischer, W.A. 1997. "The diminishing utility of the product/process matrix." *International Journal of Operations & Production Management*, Vol. 17, No. 1, pp. 65–84.

Metrix. 2021. *Additive manufacturing 3D printing year in review 2020–21*. Metrix Connect, LLC.

Mitsouras, D., George, E. and Rybicki, F.J. 2017. "Quality and safety of 3D-printed medical models." In: Rybicki, F.J. and Grant, G.T. (Eds.) *3D printing in medicine*. Cham, Switzerland: Springer.

Mitsouras, D. and Liacouras, P.C. 2017. "3D printing technologies." In: Rybicki, F.J. and Grant, G.T. (Eds.) *3D printing in medicine*. Cham, Switzerland: Springer.

Momeni, F., Liu, X. and Ni, J. 2017. "A review of 4D printing." *Materials & Design*, Vol. 122, pp. 42–79.

Moore, G.A. 2014. *Crossing the chasm*. 3rd edition. New York: HarperCollins Publishers.

Morelli, J. 2011. "Environmental sustainability: a definition for environmental professionals." *Journal of Environmental Sustainability*, Vol. 1, No. 1, article 2.

Murphy, S.V. and Atala, A. 2014. "3D bioprinting of tissues and organs." *Nature Biotechnology*, Vol. 32, No. 8, pp. 773–785.

Nadagouda, M.N., Ginn, M. and Rastogi, V. 2020, "A review of 3D printing techniques for environmental applications." *Current Opinion in Chemical Engineering*, Vol. 28, pp. 173–178.

Najmon, J.C., Raeisi, S. and Tovar, A. 2019. "Review of additive manufacturing technologies and applications in the aerospace industry." In: Froes, F. and Boyer, R. (Eds.) *Additive manufacturing for the aerospace industry*. Amsterdam: Elsevier.

Nam, Y.J., Hwang, Y.K., Park, J.W. and Lim, Y.M. 2019. "Fiber-reinforced cementitious composite design with controlled distribution and orientation fibers using three-dimensional printing technology." In: Sanjayan, J.G., Nazari, A. and Nematollahi, B. (Eds.) *3D concrete printing technology: Construction and building applications*. Oxford: Butterworth-Heinemann.

Nayyeri, P., Zareinia, K. and Bougherara, H. 2022. "Planar and nonplanar slicing algorithms for fused deposition modeling technology: a critical review." *The International Journal of Advanced Manufacturing Technology*, Vol. 119, pp. 2785–2810.

Nematollahi, B., Xia, M., Vijay, P. and Sanjayan, J.G. 2019. "Properties of extrusion-based 3D printable geopolymers for digital construction applications." In: Sanjayan, J.G., Nazari, A. and Nematollahi, B. (Eds.) *3D concrete printing technology: Construction and building applications*. Oxford: Butterworth-Heinemann.

Nerella, V.N. and Mechtherine, V. 2019. "Studying the printability of fresh concrete for formwork-free concrete onsite 3D printing technology (CONPrint3D)." In: Sanjayan, J.G., Nazari, A. and Nematollahi, B. (Eds.) *3D concrete printing technology: Construction and building applications*. Oxford: Butterworth-Heinemann.

Niinimäki, K., Peters, G., Dahlbo, H., Perry, P., Rissanen, T. and Gwilt, A. 2020. "The environmental price of fast fashion." *Nature Reviews Earth & Environment*, Vol. 1, No. 4, pp. 189–200.

Noor, N. Shapira, A., Edri, R. Gal, I., Wertheim, L. and Dvir, T. 2019. "3D printing of personalized thick and perfusable cardiac patches and hearts." *Advanced Science*, Vol. 6.

Novakov, T., Jackson, M.J., Robinson, G.M., Ahmed, W. and Phoenix, D.A. 2017. "Laser sintering of metallic medical materials – a review." *International Journal of Advanced Manufacturing Technology*, Vol. 93, pp. 2723–2752.

O'Connell, C.D., Di Bella, C., Thompson, F., Augustine, C., Beirne, S., Cornock, R., Richards, C.J., Chung, J., Gambhir, S., Yue, Z., Bourke, J., Zhang, B., Taylor, A., Quigley, A., Kapsa, R., Choong, P. and Wallace, G.G. 2016. "Development of the Biopen: a handheld device for surgical printing of adipose stem cells at a chondral wound site." *Biofabrication*, Vol. 8.

Oberti, I. and Plantamura, E. 2015. "Is 3D printed house sustainable?" Paper presented at CISBAT 2015, September 9-11, Lausanne, Switzerland.

Öblom, H., Sjöholm, E., Rautamo, M. and Sandler, N. 2019. "Towards printed pediatric medicines in hospital pharmacies: comparison of 2D and 3D-printed orodispersible warfarin films with conventional oral powders in unit dose sachets." *Pharmaceutics*, Vol. 11, pp. 314–347.

Olhager, J. 2010. "The role of the customer order decoupling point in production and supply chain management." *Computers in Industry*, Vol. 61, pp. 863–868.

Ostwald, P.F. and Muñoz, J. 1997. *Manufacturing processes and systems*. 9th edition. New York: John Wiley & Sons.

Ozbolat, I.T., Peng, W. and Ozbolat, V. 2016. "Application areas of 3D bioprinting." *Drug Discovery Today*, Vol. 21, No. 8, pp. 1257–1271.

Paasi, J., Valkokari, K., Luoma, T., Hytönen, H., Nystén-Haarala, S. and Huhtilainen, L. 2010. "Innovation management challenges of a system integrator in innovation networks." *International Journal of Innovation Management*, Vol. 14, No. 6, pp. 1047–1064.

Pandey, D., Agrawal, M. and Pandey, J.S. 2011. "Carbon footprint: current methods of estimation." *Environmental Monitoring and Assessment*, Vol. 178, No. 1, pp. 135–160.

Pant, A., Lee, A.Y., Karyappa, R., Lee, C.P., An, J., Hashimoto, M., Tan, U.-X., Wong, G., Chu, C.K. and Zhang, Y. 2021. "3D food printing of fresh vegetables using food hydrocolloids for dysphagic patients." *Food Hydrocolloids*, Vol. 114.

Parr, W.C.H., Burnard, J.L., Wilson, P.J. and Mobbs, R.J. 2019. "3D printed anatomical (bio)models in spine surgery: clinical benefits and value to health care providers." *Journal of Spine Surgery*, Vol. 5, No. 4, pp. 549–560.

Parvez, M., Ullah, N., Sabuj, M.A. and Islam, S. 2018. "Profit maximization of Dell Inc. through build-to-order supply chain for laptop manufacturing." *American Journal of Industrial and Business Management*, Vol. 8, pp. 1657–1671.

Pascal, V.J., Sprott, D.E. and Muehling, D.D. 2002. "The influence of evoked nostalgia on consumers' responses to advertising: an exploratory study." *Journal of Current Issues & Research in Advertising*, Vol. 24, No. 1, pp. 39–47.

Pearson, H.A. and Dubé, A.K. 2022. "3D printing as an educational technology: theoretical perspectives, learning outcomes, and recommendations for practice." *Education and Information Technologies*, pp. 1–28.

Pereira, A.C. and Romero, F. 2017. "A review of the meanings and the implications of the Industry 4.0 concept." *Procedia Manufacturing*, Vol. 13, pp. 1206–1214.

Perkins, I. and Skitmore, M. 2015. "Three-dimensional printing in the construction industry: a review." *International Journal of Construction Management*, Vol. 15, No. 1, pp. 1–9.

Perrot, A. (Ed.) 2019. *3D printing of concrete*. Hoboken: John Wiley & Sons.

Perry, A. 2018. "3D-printed apparel and 3D-printer: exploring advantages, concerns, and purchases." *International Journal of Fashion Design, Technology and Education*, Vol. 11, No. 1, pp. 95–103.

Platts, K.W. and Song, N. 2010. "Overseas sourcing decisions – the total cost of sourcing from China." *Supply Chain Management, An International Journal*, Vol. 15, No. 4, pp. 320–331.

Porter, M.E. 1980. *Competitive strategy: Techniques for analyzing industries and competitors*. New York: The Free Press.

Porter, M.E. 1985. *Competitive advantage: Creating and sustaining superior performance*. New York: The Free Press.

Porter, M.E. and Stern, S. 2001. "Innovation: location matters." *MIT Sloan Management Review*, Summer, pp. 28–36.

Prăvălie, R. 2018. "Major perturbations in the Earth's forest ecosystems. Possible implications for global warming." *Earth-Science Reviews*, Vol. 185, pp. 544–571.

Ravi, P., Chepelev, L.L., Stichweh, G.V., Jones, B.S. and Rybicki, F.J. 2022. "Medical 3D printing dimensional accuracy for multi-pathological anatomical models 3D printed using MEX." *Journal of Digital Imaging*. https://doi.org/10.1007/s10278-022-00614-x

Rayna, T. and Striukova, L. 2016. "From rapid prototyping to home fabrication: How 3D printing is changing business model innovation." *Technological Forecasting and Social Change*, Vol. 102, No. 1, pp 214–224.

Rayna, T. and Striukova, L. 2021. "Assessing the effect of 3D printing technologies on entrepreneurship: an exploratory study." *Technological Forecasting and Social Change*, Vol. 164, pp. 120483.

Rayna, T., Striukova, L. and Darlington, J. 2015. "Co-creation and user innovation: the role of online 3D printing platforms." *Journal of Engineering and Technology Management*, Vol. 37, pp. 90–102.

Redwood, B., Schöffer, F. and Garret, B. 2017. *The 3D printing handbook*. Amsterdam: 3D Hubs B.V.

Ren, J., Zhang, Y. and Zhao, D. et al. 2022. "Strong yet ductile nanolamellar high-entropy alloys by additive manufacturing." *Nature*, Vol. 608, pp. 62–68.

Rideout, B. 2011. "Printing the impossible triangle: the copyright implications of three-dimensional printing." *The Journal of Business, Entrepreneurship & the Law*, Vol. 5, No. 1, pp. 161–177.

Rogers, E.M. 1995. *Diffusion of innovations*. 4th edition. New York: The Free Press.

Rouhana, C.M., Aoun, M.S., Faek, F.S., Eljazzar, M.S. and Hamzeh, F.R. 2014, "The reduction of construction duration by implementing contour crafting (3D printing)", *Proceedings IGLC*, 22 June, Oslo, Norway, pp. 1031–1042.

Rucker, R.R., Thurman, W.N. and Burgett, M. 2019. "Colony collapse and the consequences of bee disease: market adaptation to environmental change." *Journal of the Association of Environmental and Resource Economists*, Vol. 6, No. 5, pp. 927–960.

Rudberg, M. and Wikner, J. 2004. "Mass customization in terms of the customer order decoupling point." *Production Planning and Control*, Vol. 15, No. 4, pp. 445–458.

Russell, R., Wells, D., Poorganji, B., Ott, E., Nakagawa, T., Sandoval, H., Shamsaei, N. and Seifi, M. 2019. "Qualification and certification of metal additive manufactured hardware for aerospace applications." In: Froes, F. and Boyer, R. (Eds.) *Additive manufacturing for the aerospace industry*. Amsterdam: Elsevier.

Rybicki, F.J. and Grant, G.T. (Eds.) 2017. *3D printing in medicine*. Cham, Switzerland: Springer.

Saccone, M.A., Gallivan, R.A., Narita, K., Yee, D.W. and Greer, J.R. 2022. "Additive manufacturing of micro-architected metals via hydrogel infusion." *Nature*, Vol. 612, pp. 685–690.

Salmi, M. 2021. "Additive manufacturing processes in medical applications." *Materials*, Vol. 14, pp. 191–207.

Salmi, M. et al. 2018. "Digital spare parts.". Published by Aalto University and VTT Technical Research Centre of Finland Ltd. This report is available at: http://urn.fi/URN. ISBN:978-952-60-3746-2.

Sampson, S.E. 2001. *Understanding service businesses: Applying principles of the unified services theory*. John Wiley & Sons.

Sampson, S.E. and Froehle, C.M. 2006. "Foundations and implications of a proposed unified services theory." *Production and Operations Management*, Vol. 15, No. 2, pp. 329–343.

Sandström, C.G. 2016. "The non-disruptive emergence of an ecosystem for 3D printing – insights from the hearing aid industry's transition 1989–2008." *Technological Forecasting & Social Change*, Vol. 102, pp. 160–168.

Sanguinetti, P., Almazam, K., Humaidan, O. and Colistra, J. 2019. "Evaluating the Potential of High Performance Concrete 3D-Printing for Zero Energy Homes". In *16th International Conference of the International Building Performance Simulation Association, Building Simulation 2019* (pp. 5068–5075). International Building Performance Simulation Association.

Sanicola, H.W., Stewart, C.E., Mueller, M., Ahmadi, F., Wang, D., Powell, S.K., Sarkar, K., Cutbush, K., Woodruff, M.A. and Brafman, D.A. 2020. "Guidelines for establishing a 3-D printing biofabrication laboratory." *Biotechnology Advances*, Vol. 45.

Sanjayan, J.G. and Nematollahi, B. 2019. "3D Concrete printing for construction applications." In: Sanjayan, J.G., Nazari, A. and Nematollahi, B. (Eds.) *3D concrete printing technology: Construction and building applications*. Oxford: Butterworth-Heinemann.

Sansone, C., Hilletofth, P. and Eriksson, D. 2017. "Critical operations capabilities for competitive manufacturing: a systematic review." *Industrial Management & Data Systems*, Vol. 117, No. 5, pp. 801–837.

Sansone, C., Hilletofth, P. and Eriksson, D. 2020. "Evaluation of critical operations capabilities for competitive manufacturing in a high-cost environment." *Journal of Global Operations and Strategic Sourcing*, Vol. 13, No. 3, pp. 229–250.

Santana, L., Ahrens, C.H., da Costa Sabino Netto, A. and Bonin, C. 2017. "Evaluating the deposition quality of parts produced by an open-source 3D printer." *Rapid Prototyping Journal*, Vol. 23, No. 4, pp. 796–803.

Santos, E.O., Oliveira, P.L.E., de Mello, T.P., dos Santos, A.L.S., Elias, C.N., Choi, S.H. and de Castro, A.C.R. 2022. "Surface characteristics and microbiological analysis of a vat-photopolymerization additive-manufacturing dental resin." *Materials*, Vol. 15, pp. 425–438.

Schwab, K. 2016. *The fourth industrial revolution*. Geneva: World Economic Forum.

Shafer, S.M., Smith, H.J. and Linder, J.C. 2005. "The power of business models." *Business Horizons*, Vol. 48, pp. 199–207.

Shao, L. et al. 2022. "A chemical approach for the future of PLA upcycling: from plastic wastes to new 3D printing materials." *Green Chemistry*. Vol. 24, no 22, pp. 8716–8724. Advance article.

Sharif, N.M. 1988. "Basis for techno-economic policy analysis." *Science and Public Policy*, Vol. 15, No. 4, pp. 217–229.

Sheikh, A., Chepelev, L., Christensen, A.M., Mitsouras, D., Schwarz, B.A. and Rybicki, F. 2017. "Beginning and developing a radiology-based in-hospital 3D printing lab." In: Rybicki, F.J. and Grant, G.T. (Eds.) *3D printing in medicine*. Cham, Switzerland: Springer.

Shi, Y. and Herniman, J. 2023. "The role of expectation in innovation evolution: exploring hype cycles". *Technovation*, Vol. 119, pp. 102459.

Shuaib, M., Haleem, A., Kumar, S. and Javaid, M. 2021. "Impact of 3D printing on the environment: a literature-based study." *Sustainable Operations and Computers*, Vol. 2, pp. 57–63.

Singh, S.N., Venkatesh, V.S.S. and Deoghare, A.B. 2021. "A review on the role of 3D printing in the fight against COVID-19: safety and challenges." *Rapid Prototyping Journal*, Vol. 27, No. 2, pp. 407–420.

Singhal, S. et al. 2020. "3D printing foods: paving way towards novel foods." *Annals of the Brazilian Academy of Sciences*, Vol. 92, No. 3.

Skardal, A. 2018. "Principles and applications of bioprinting." In: Khademhosseini, A. and Camci-Unal, G. (Eds.) *3D bioprinting in regenerative engineering: Principles and applications*. Boca Raton: CRC Press.

Slack, N. and Lewis, M. 2011. *Operations strategy*, 3rd edition. Harlow, UK: Pearson Education Limited.

Sole, J. et al. 2017. *The need for speed hits athletic wear*. New York: Morgan Stanley.

Sonobi, M., Amziane, S. and Perrot, A. 2019. "Mechanical behavior of 3D printed cement materials." In: Perrot, A. (Ed.) *3D printing of concrete*. Hoboken: John Wiley & Sons.

Srai, J.S. et al. 2016. "Distributed manufacturing: scope, challenges and opportunities." *International Journal of Production Research*, Vol. 54, No. 23, pp. 6917–6935.

Stalk, G. Jr. and Hout, T.M. 1990. *Competing against time: How time-based competition is reshaping global markets*. New York: The Free Press.

Stearns, P.N. 2013. *The industrial revolution in world history*. 4th edition. Boulder: Westview Press.

Steenhuis, H.J. 2015. *International operations: How multiple international environments impact productivity and location decisions*. New York: Business Expert Press.

Steenhuis, H.J. and de Bruijn, E.J. 2006. "High technology revisited: definition and position." In: Chai, H.C., Hang, C.C. and Xie, M., *IEEE international conference on management of innovation and technology*, Vol. 2, pp. 21–23 June, Singapore.

Steenhuis, H.J. and de Bruijn, E.J. 2012. "Technology and economic development: a literature review." *International Journal of Innovation and Technology Management*, Vol. 9, No. 5.

Steenhuis, H.J., Fang, X. and Ulusemre, T. 2020. "Global diffusion of innovation during the fourth industrial revolution: the case of additive manufacturing or 3D printing." *International Journal of Innovation and Technology Management*, Vol. 17, No. 1.

Steenhuis, H.J., Fang, X. and Ulusemre, T. 2022. "High technology, performance and price: the case of additive manufacturing through desktop 3D printers." *Technology Analysis & Strategic Management*. https://www.tandfonline.com/doi/full/10.1080/09537325.2022.2090330

Steenhuis, H.J. and Pretorius, L. 2017. "The additive manufacturing innovation: a range of implications." *Journal of Manufacturing Technology Management*, Vol. 28, No. 1, pp. 122–143.

Suiker, A.S.J. 2018. "Mechanical performance of wall structures in 3D printing processes: theory, design tools and experiments." *International Journal of Mechanical Sciences*, Vol. 137, pp. 145–170.

Sun, W., Starly, B., Daly, A.C., Burdick, J.A., Groll, J., Skeldon, G., Shu, W., Sakai, Y., Shinohara, M., Nishikawa, M., Jang, J., Cho, D.W., Nie, M., Takeuchi, S., Ostrovidov, S., Khademhosseini, A, Kamm, R.D., Mironov, V., Moroni, L. and Ozbolat, I.T. 2020. "The bioprinting roadmap." *Biofabriation*, Vol. 12.

Sutton, P. 2004. *A Perspective on environmental sustainability? A paper for the Victorian Commissioner for Environmental Sustainability*, pp. 1–32.

Tabari, H. 2020. "Climate change impact on flood and extreme precipitation increases with water availability." *Scientific Reports*, Vol. 10, No. 1, pp. 1–10.

Tarrade, T., Doucet, F., Saint-Lô, N., Llari, M. and Behr, M. 2019. "Are custom-made foot orthoses of any interest on the treatment of foot pain for prolonged standing workers?" *Applied Ergonomics*, Vol. 80, pp. 130–135.

Thomas, V.J. and Maine, E. 2019. "Market entry strategies for electric vehicle start-ups in the automotive industry – lessons from Tesla Motors." *Journal of Cleaner Production*, Vol. 235, pp. 653–663.

Tidd, J. and Bessant, J. 2009. *Managing innovation: Integrating technological, market and organizational change*, 4th edition. West Sussex: John Wiley & Sons.

Tornatzky, L.G. and Fleischer, M. 1990. *The process of technological innovation*. Lexington: Lexington Books.

Truxova, V., Safka, J., Seidl, M., Kovalenko, I., Volesky, L. and Ackermann, M. 2020. "Ceramic 3D printing: comparison of SLA and DLP technologies." *MM Science Journal*, June, pp. 3905–3911.

Tyagi, S., Yadav, A. and Deshmukh, S. 2022. "Review of mechanical characterization of 3D printed parts created using material jetting process." *Materials Today: Proceedings*, Vol. 51, pp. 1012–1016.

Üçgül, M. and Altıok, S. 2023. "The perceptions of prospective ICT teachers towards the integration of 3D printing into education and their views on the 3D modeling and printing course." *Education and Information Technologies*, pp. 1–31. https://link.springer.com/article/10.1007/s10639-023-11593-z#Abs1

Ullah, A., Khan, D., Khan, I. and Zheng, S. 2018. "Does agricultural ecosystem cause environmental pollution in Pakistan? Promise and menace." *Environmental Science and Pollution Research*, Vol. 25, pp. 13938–13955.

Urbanic, R.J., Kilani, M. and Hassoun, A. 2013. "Targeted reverse engineering techniques for generating architectural solid models for additive manufacturing fabrication." *Computer-Aided Design & Applications*, Vol. 10, No. 4, pp. 585–602.

Vanderploeg, A., Lee, S.E. and Mamp, M. 2017. "The application of 3D printing technology in the fashion industry." *International Journal of Fashion Design, Technology and Education*, Vol. 10, No. 2, pp. 170–179.

VanEngelsdorp, D., Evans, J.D., Saegerman, C., Mullin, C., Haubruge, E., Nguyen, B.K., Frazier, M., Frazier, J., Cox-Foster, D., Chen, Y. and Underwood, R. 2009. "Colony collapse disorder: a descriptive study." *PloS one*, Vol. 4, No. 8. Article e6481.

Veld, J. in't 1992. *Analyse van organisatie problemen*. Leiden, the Netherlands: Stenfert Kroese Uitgevers.

Venios, X., Korkas, E., Nisiotou, A. and Banilas, G. 2020. "Grapevine responses to heat stress and global warming." *Plants*, Vol. 9, No. 12, p. 1754.

Venkatesh, V., Morris, M.G., Davis, G.B. and Davis, F.D. 2003. "User acceptance of information technology: toward a unified view." *MIS Quarterly*. Vol 27, No. 3, pp. 425–478.

Verboeket, V. et al. 2021. "Additive manufacturing for localized medical parts production: a case study." *IEEE Access*, Vol. 9, pp. 25818–25834.

Vereecke, A. and Van Dierdonck, R. 2002. "The strategic role of the plant: testing Ferdow's model." *International Journal of Operations and Production Management*, Vol. 22, No. 5, pp. 492–514.

Verma, V.K., Kamble, S.S., Ganapathy, L. Belhadi, A. and Gupta, S. 2022. "3D Printing for sustainable food supply chains: modelling the implementation barriers." *International Journal of Logistics: Research and Applications*. DOI: 10.1080/13675567.2022.2037125

Vicente-Serrano, S.M., Quiring, S.M., Pena-Gallardo, M., Yuan, S. and Dominguez-Castro, F. 2020. "A review of environmental droughts: increased risk under global warming?" *Earth-Science Reviews*, Vol. 201, p. 102953.

Virtanen, P.K., Siragusa, L. and Guttorm, H. 2020. "Introduction: toward more inclusive definitions of sustainability." *Current Opinion in Environmental Sustainability*, Vol. 43, pp. 77–82.

Vlot, A. 2004. *Glare history of the development of a new aircraft material*. New York: Kluwer Academic Publishers.

Voet, V.S. et al. 2018. "Biobased acrylate photocurable resin formulation for stereolithography 3D printing." *ACS Omega*, Vol. 3, No. 2, pp. 1403–1408.

von Tunzelman, N. 2003. "Historical coevolution of governance and technology in the industrial revolutions." *Structural Change and Economic Dynamics*, Vol. 14, pp. 365–384.

Voon, S.L., An, J., Wong, G., Zhang, Y. and Chua, C.K. 2019. "3D food printing: a categorized review of inks and their development." *Virtual and Physical Prototyping*, Vol. 14, No. 3, pp. 203–218.

Wang, C., Tana, X.P., Tora, S.B. and Lim, C.S. 2020. "Machine learning in additive manufacturing: State-of-the-art and perspectives." *Additive Manufacturing*, Vol. 36.

Wang, J.C., Dommati, H. and Hsieh, S.J. 2019. "Review of additive manufacturing methods for high-performance ceramic materials." *The International Journal of Advanced Manufacturing Technology*, Vol. 103, pp. 2627–2647.

Warren, P., Raju, N., Ebrahimi, H., Krsmanovic, M., Raghavan, S., Kapat, J. and Gosh, R. 2022. "Effect of sintering temperature on microstructure and mechanical properties of molded Martian and Lunar regolith." *Ceramics International*, Vol. 48, No. 23, Part B, pp. 35825–35833.

Warren, R., Price, J., Graham, E., Forstenhaeusler, N. and VanDerWal, J. 2018. "The projected effect on insects, vertebrates, and plants of limiting global warming to 1.5 C rather than 2 C." *Science*, Vol. 360, No. 6390, pp. 791–795.

Wittbrodt, B.T., Glover, A.G., Laureto, J., Anzalone, G.C., Oppliger, D., Irwin, J.L. and Pearce, J.M. 2013. "Life-cycle economic analysis of distributed manufacturing with open-source 3-D printers." *Mechatronics*, Vol. 23, No. 6, pp. 713–726.

Wittwer, M. and Seita, M. 2022. "A machine learning approach to map crystal orientation by optical microscopy." *npj Computational Materials*, Vol 8, No. 8.

Wohlers Associates 2017. *Wohlers Report 2017, 3D printing and additive manufacturing state of the industry, Annual worldwide progress report.* Fort Collins, CO: Wohlers Associates, Inc.

Wohlers Associates, 2020. *Wohlers Report 2020, 3D printing and additive manufacturing, Global state of the industry.* Fort Collins, CO: Wohlers Associates, Inc.

Wolf-Powers, L., Doussard, M., Schrock, G., Heying, C., Eisenburger, M. and Marotta, S. 2017. "The Maker movement and urban economic development." *Journal of the American Planning Association*, Vol. 83, No. 4, pp. 365–376.

Womack, J.P., Jones, D.T. and Roos, D. 1990. *The machine that changed the world: The story of lean production*, New York: Harper Perennial.

World Health Organization 2021. *Global expenditure on health: Public spending on the rise?* Global report. Geneva: World Health Organization.

Wright, C. 2022. *Dosimetric and toxicological analysis of 3D printer emitted particles: Final report.* Marietta: Chemical Insights.

Wu, W. and Skye, H.M. 2021. "Residential net-zero energy buildings: review and perspective." *Renewable and Sustainable Energy Reviews*, Vol. 142, article 110859.

Wüthrich, M., Gubser, M., Elspass, W.J. and Jaeger, C. 2021. "A novel slicing strategy to print overhangs without support material." *Applied Sciences*, Vol. 11, No. 18, pp. 8760.

Xia, M., Nematollahi, B. and Sanjayan, J.G. 2019a. "Development of powder-based 3D concrete printing using geopolymers." In: Sanjayan, J.G., Nazari, A. and Nematollahi, B. (Eds.) *3D concrete printing technology: Construction and building applications.* Oxford: Butterworth-Heinemann.

Xia, M., Nematollahi, B. and Sanjayan, J.G. 2019b. "Properties of powder-based 3D printed geopolymers." In: Sanjayan, J.G., Nazari, A. and Nematollahi, B. (Eds.) *3D concrete printing technology: Construction and building applications.* Oxford: Butterworth-Heinemann.

Yadid, M. et al. 2022. "Bioengineering approaches to treat the failing heart: from cell biology to 3D printing." *Nature Reviews Cardiology*, Vol. 19, No. 2, pp. 83–99.

Yang, F., Zhang, M. and Bhandari, B. 2017. "Recent developments in 3D food printing." *Critical Reviews in Food Science and Nutrition*, Vol. 57, No. 4, pp. 3145–3153.

Ye, Z., Liu, C., Tian, W. and Kan, C. 2021. "In-situ point cloud fusion for layer-wise monitoring of additive manufacturing." *Journal of Manufacturing Systems*, 61, pp. 210–222.

Zeidler, H., Klemm, D., Böttger-Hiller, F., Fritsch, S., Le Guen, M.J. and Singamneni, S. 2018. "3D printing of biodegradable parts using renewable biobased materials." *Procedia Manufacturing*, Vol. 21, pp. 117–124.

Zhu, Y., Zhang, K., Meng, Z., Zhang, K., Hodgson, P., Birbilis, N., Weyland, M., Fraser, H.L., Lim, S.C.V., Peng, H. and Yang, R. 2022. "Ultrastrong nano-twinned titanium alloys through additive manufacturing." *Nature Materials*, Vol. 21, No. 11, pp. 1258–1262.

Index

Pages in *italics* refer figures and pages in **bold** refer tables.

Companies producing AM or related technology

Companies applying or working with AM technology

9781032505725